## Building Research: Design, Construction and Technologies

**Series Editor**

Bárbara Rangel, Civil Engineering, University of Porto, Porto, Portugal

*Building Research: Design, Construction and Technologies* brings together knowledge from civil engineering and architecture to make an interdisciplinary analysis of a recognized building/project or a building construction research. In each volume, the topic is a building or an aspect of a building that catalyzes the contributions of invited authors in their fields of specialization, with professional and academic backgrounds. To make the bridge between the scientific research and the construction site problems a parallel reading of the working development and the technological issues raised by that construction problem are presented. Authors, architects and engineers are interviewed and analyze the different projects in distinct stages, from concept to construction drawings, following the development of the building's design and construction process. The series treats topics such as building technology, construction management, acoustics, maintenance, prefabrication, amongst others. As a complementary objective, authors with different backgrounds: Engineers and Architects, Researchers and Designers are invited to to contribute to the understanding of specific buildings through the analysis of those issues that influenced the development of the design or that appeared during the construction or the facility management phases.

More information about this series at http://www.springer.com/series/13572

Bárbara Rangel
José Manuel Amorim Faria • Vitor Abrantes
Editors

# Museum Technology and Architecture

Springer

*Editors*
Bárbara Rangel
Civil Engineering
Universidade do Porto
Porto, Portugal

José Manuel Amorim Faria
Civil Engineering
Universidade do Porto
Porto, Portugal

Vitor Abrantes
Civil Engineering
Universidade do Porto
Porto, Portugal

ISSN 2366-6307 ISSN 2366-6315 (electronic)
Building Research: Design, Construction and Technologies
ISBN 978-3-319-76170-1 ISBN 978-3-319-76172-5 (eBook)
https://doi.org/10.1007/978-3-319-76172-5

Library of Congress Control Number: 2018964068



This Springer imprint is published by the registered company Springer Nature Switzerland AG.
The registered company address is: Gewerbestrasse 11, 6330 Cham, Switzerland

# Preface

This is the second issue of the book series on building construction *Building Research: Design, Construction and Technologies* published by Springer and prepared and edited at the University of Porto in Portugal, under the theme *Museum Technology and Architecture*. Two different museums are described and scrutinized using construction technology as the common language that brings Architecture and Engineering together. Those two buildings are the Museum of the Ibere Camargo Foundation at Porto Alegre, Rio Grande do Sul, Brazil, and the Coach Museum in Lisbon, Portugal. The Ibere Cargo Foundation Museum, designed by Alvaro Siza, was inaugurated in 2008, after 8 years of design and 4 years of construction in very close and strong cooperation with engineers Jorge Silva from GOP Engineering and Raul Bessa of GET, both design companies from Oporto, Portugal. The Coach Museum in Lisbon, designed by the architects Paulo Mendes da Rocha and Ricardo Bak Gordon, with Rui Furtado leading the engineering team of AFA consult, was inaugurated in 2015 after 2 years of design, 4 years of construction, and 2 years of preparation for the opening. Both projects had a special emphasis on the design process as a construction language, achieved through a close collaboration promoted by the integrated design methodology that both teams follow. Besides their importance from architectural and urban points of view, these two buildings suggest interesting topics that are relevant to current building research, such as sustainability in general, the construction of facades with a heavy use of unrendered white concrete, and the integration of all of the technical infrastructure required to build a successful high-tech museum. Both designs aimed to produce state-of-the-art sustainable museums, which may be considered as an objective fulfilled.

The book is divided into two different parts, each one related to one of the two museums. Each part presents an interview (chapters "Interview with Álvaro Siza" and "Display of Technique", Parts I and II) with the architect and the team that was responsible for the design of each museum and four articles (chapters "Iberê Carmargo Foundation: Why Was This a Successful Project?", "Hygrothermal Behaviour Control in the Crawlspaces of Historical Museums", "The Efficiency of Air Conditioning Energy Systems in the Iberê Camargo Contemporary Art Museum – Rio Grande Do Sul – Brazil", "The Engineering of the New Coach Museum", "Lighting and Museums", "Museums Seen Through the Lens of Building Physics", "From the Utilisation of Rainwater to the Re-use of Grey Water: Some Thoughts Concerning the Iberê Camargo Foundation Museum", and "Coordination of Work on the Coach Museum Supported by BIM Methodology", Parts I and II) that explore the scientific and conceptual issues resulting from the project analyses. In both parts, the project drawings, conceptual sketches, and images of the building that illustrate the understanding of the analyses of each project are presented.

The first building, in Porto Alegre, Brazil, is a huge success and has become a strong icon of that city, the capital of the most southern state of Brazil, Rio Grande do Sul. It has changed the landscape of that specific place and heavily connects the building and the Ibere Camargo art collection with the lagoon in front of the building. The interview with Alvaro Siza is titled "Construction as a Conceptual Language." Many issues are discussed under that umbrella, following a route that finishes on an "intimate" discussion of the processes of design, research, and experimentation on materials and construction systems and solutions, finally landing on the subsequent technical assistance involved in the construction itself.

José Luiz Canal, the Brazilian project manager of this construction, very passionately and thoroughly describes the construction phase in chapter 2 of Part I, "Iberê Carmargo Foundation: Why Was This a Successful Project?" Particularly interesting in this chapter is the description of the white concrete production process, including the formwork, and of the participation of local academics who gave advice and helped to perform the quality control tasks.

In chapter "The Efficiency of Air Conditioning Energy Systems in the Iberê Camargo Contemporary Art Museum – Rio Grande Do Sul – Brazil" of Part I, the efficiency of the heating, ventilation, and air-conditioning (HVAC) energy systems in the Ibere Camargo Museum is described and synthetically analyzed. The building is a fine example of a low energy consumption building. Its design demonstrates a solution that uses the energy to support its HVAC installations in a very modern, environmentally friendly, and economic way.

Chapter "Lighting and Museums" of Part I addresses lighting in museums. The lighting solutions in museums must be completely adapted to the architectural design and philosophy, with a focus on the main purpose of the lighting system in any museum, which is to illuminate the museum's works of art with a color temperature and an intensity suitable for each purpose and exhibition room. In general, there should be controlled, diffused natural light adjusted to the architectural concept. The author of the chapter presents a very useful and detailed technical analysis of the problem, illustrated with real examples that include the Ibere Camargo Museum.

Following the main objective of all of the engineering designs of the Ibere Camargo museum, which was to build a "state-of-the-art sustainable museum," chapter "From the Utilisation of Rainwater to the Re-use of Grey Water: Some Thoughts Concerning the Iberê Camargo Foundation Museum" of Part I presents the case of the utilization of rainwater and the reuse of gray water that was considered and applied in this specific museum. Actually, water efficiency measures, such as the two mentioned above, represent a strategic necessity, given the risk of shortages of water, especially in countries or regions with relatively low rain precipitation, which is the case with Rio Grande do Sul in Brazil, although this is not the case in all of the other Brazilian states.

Part II of the book is dedicated to the new Coach Museum in Lisbon, Portugal, designed by the Brazilian architect Paulo Mendes da Rocha. This museum today represents one very good example of the renovation process that Lisbon has been experiencing in more recent years and constitutes a very important new attraction of the highly acclaimed touristic zone where it was built (Belém). The interview with Paulo Mendes da Rocha is titled "Display of Technique." This refers to the fact that this museum follows the actual trends of museums that are "easy-to-use interactive learning environments," considerably focused on a very specific subject or objects on display. The interview more widely covers the career of Paulo Mendes da Rocha but ultimately concentrates on the specific design of this museum.

Chapter "Hygrothermal Behaviour Control in the Crawlspaces of Historical Museums" of Part II synthetically presents the engineering of the building and its main designers. The chapter presents all of the engineering designs in an integrated form, emphasizing the relations among them, as well as that with the "architectural concept" that is also described. Quoting its authors "… (this museum) is a machine whose efficiency and success depends as much on how its various components fulfill their function (…) as in its ability to contain spaces capable of surprising and inspiring."

Chapter "The Engineering of the New Coach Museum" of Part II concerns building physics and museums. The author, who is not one of the engineers from the design team but is a widely known and very reputed academic professor, has produced a short and very easy-to-use "tutorial" of the problem of the physics of building in a museum. Most buildings are built to house people. Museums are different! The artifacts stored, preserved, and displayed, and not the visitors, determine what the indoor environment should look like.

Chapter "Museums Seen Through the Lens of Building Physics" of Part II is dedicated to BIM methodology and the positive role that this technology played in the coordination of the different designs and in the preparation of construction work by the contractor. The new Coach Museum in Lisbon has a dense steel structure, and the architectural design aimed to provide broad and free spaces/volumes for the display of the coaches. In this context, it is important to stress the challenge of achieving compatibility among all of the engineering specialties with the placement of hundreds of meters of pipes, cables, and tubes in a confined space above the false ceiling built under the steel structure roof of the building. Considering that, it was decided by the architect that all of the various services should remain on view above a nonopaque metal grating. The design of all of those services should represent order and be, in itself, a "display of technology" parallel to the floor exhibition of coaches. BIM was essential for fulfilling this objective, and the decision to use it was made by the contractor, as the positive effect described in the chapter evidences.

Finally, chapter "Coordination of Work on the Coach Museum Supported by BIM Methodology" of Part II closes the book by focusing on the need to control humidity and temperature in a museum with special care, so that the displayed works of art are well preserved. As Prof. Hugo Hens points out (see chapter "The Engineering of the New Coach Museum", Part II), "…some architects too often overlook these building physics-related aspects and design museums that demand corrective measures afterwards, once built."

This book aims therefore to continue narrowing the gaps between architects and state-of-the-art technology using construction systems and ordinary knowledge of physics as a common language that all construction stakeholders should practice and widely understand.

Porto, Portugal
4 Jan 2018

José Manuel Amorim Faria
Bárbara Rangel

# Contents

# Part I

# Ibere Camargo Foundation

Project by:
Architect Álvaro Siza
GOP Engineering
GET AVAC Engineering
Raul Serafim Electrical Engineering

# The Building Master

Bárbara Rangel, José Manuel Amorim Faria,
and João Pedro Poças Martins

**Building Research (BR)** – In this interview, we would like to explore how building systems and technology enter into the language of your work, particularly in the *Iberê Camargo Foundation building*, a focus of this magazine. What's noticeable in the course of your work, from the *Swimming Pool in Leça* to the *Iberê Camargo Foundation building*, is the search for an artistic expression through building technologies, At first, looking for plasticity of form through the construction materials, like in the *Marés Swimming Pool*, or the *Casa de Chá*. In *Bouça*, using construction as an experimental technology, realising the capabilities of resistant masonry or even the handcrafted cappotto in the *Carlos Ramos Pavilion*. Now, you seem to have found a 'constructive language' that allows for greater freedom to do what you sought to do all along: treat the space as a sculpture.

Is this the route?

Is the architectural space a sculptural space?

**AS** – The decision varies a lot. It depends on the context, it depends on the maintenance problems and it depends on the evolution of the idea, of the design. There is, however, a crossover of information, partly indirectly related to technology, and partly directly related to it. First, one must understand the environment we are working in, which you can do easily. It wouldn't be natural to choose concrete for a job in a village, because you would surely quickly understand that it would be difficult to do. This problem has even more variables.

Le Corbusier created one of the greatest masterpieces of architecture using the worst concrete in the world. He took advantage of the difficulty of implementing and adopting a certain technology on that site, he broke the rules and achieved absolutely wonderful results.

In my work, there is a fundamental point, from the start of the project – since I am talking in the presence of two engineers – the involvement of engineers of various specialities. I can't do without them! The way I work must include engineers from various specialized fields from the beginning of the design process.

Nevertheless, if the project is simple, the previous experience of the architect will be enough to start the designing process. In the most common cases, the choices to be made weigh heavily. In the case of *Iberê Camargo*, I hadn't visited Porto Alegre previous to the commissioning of the design, but I did have documentation, photographs, etc. So, the first studies had a lot to do with the landscape, with photography. It was an extremely difficult location. At first sight, it seemed impossible. Narrow, and decreasing in a triangle towards the slope. The first survey was mainly done through sketches. It was like a stroll around the various hypotheses. We navigated quickly through a lot of hypotheses at first.

When I started to firm up on an idea, it immediately tended towards a building in unfinished concrete. Of course, this also had to do with my own recent experiences, and with the tradition of concrete in Brazil, in Brazilian modernism ... It had to do with many things. In architecture, things never have just one reason, there are always various reasons. You have to establish relationships, analyse many things, to arrive at something that conveys the conviction of what you are doing… That projects the conviction of what you are doing…

**BR** – The way you work the raw material of the construction of buildings is always very plastic, very sculpted, be it a building in granite or in concrete.

In recent works, you seem to have found the same 'melody' that you then adapt to different sonatas/programmes. *Serralves* has the same details as *Iberê Camargo*, but they are very distinct buildings.

Does the search for the languages or this constructive typology offer you more freedom to sculpt the space?

**AS** – No, I don't think so.

I think there is, on the one hand – why not say it? – a certain complacency in using what has already been mastered, or what you think you've mastered. In using that which one knows better, you are responding to the difficulty of getting

B. Rangel (✉) · J. M. Amorim Faria · J. P. Poças Martins
Universidade do Porto, Porto, Portugal
e-mail: brangel@fe.up.pt



B. Rangel et al. (eds.), *Museum Technology and Architecture*, Building Research: Design, Construction and Technologies,
https://doi.org/10.1007/978-3-319-76172-5_1

a good result in terms of execution. I don't have a general rule; it depends very much on the circumstances. The progress of a project depends very much on the client, of course, because it's him that's paying. It depends a lot, too, on the builder (although this will be more difficult to predict). In the case of the *Iberê Camargo Foundation*, it depended a lot (I'll explain why later).

But usually, you don't know who the builder will be, you don't even have a say in choosing him. You have less and less involvement. At most, one can say 'beware.' The decision is not ours... and we are not even consulted. This has to do with the evolution of the status of architects, to the way things stand now. At the moment, I'm doing a building in Switzerland, in Basel. It's a collection of large laboratories, for which the developer/client demanded from the start that all the buildings should be made out of glass, transparent. Something that I'd never done. But actually, they are never transparent.

**BR** – You had done the project on Avenida da Ponte…

**AS** – Yes, but it was never built, for very precise reasons. But I did the project for this building in Switzerland. It's a reinforced concrete structure, pillars, beams, a cage. Then, it's finished in glass, taking the necessary care with thermal insulation, protection of privacy, etc., etc. The solutions vary a lot, you can't have a fixed rule. If the rules were fixed, it would be very boring, always the same thing.

**BR** – This question of technological experimentation that started with resistant masonry, the concept and moment of reinforcement, and now the question of "white concrete." Do you think there is, in the course of your work, a constructive language with the materials?

**AS** – Yes, there's always a language. It could be a cacophonous language or a polite language. Of course, the fact that a particular type of structure is adopted, a certain material, coatings, etc., will differentiate the building. At the moment of choosing these options, what I'm most concerned with is the context. I'm concerned the eventual different results. Working in an historical district can mean respecting the materials, the spirit of what is there, transposed to the present day. But it can also be exactly the opposite. The option for one or the other has to be very calculated. If there's a call for rupture, for fracture, or if the work does not justify it.

It can't just be according to the whim of the architect: 'Something brilliant would go down well here, cutting through all this'…No, we need other things – not the design or the architecture – calling for or leading to a rupture, or to continuity.

**BR** – This is evident in the project you did for *Avenida da Ponte* in the '60s, and then for the *Museu da Cidade*, just across the street. A much more integrated, much more sober work.

What changed from one to the other?

**AS** – First, a brief historical explanation. The first Avenida da Ponte was made as part of a major transformation of that area, spearheaded by Robert Auzelle, who was then the City Planning Officer for Porto. He asked for the implementation of conditions for mobility, easy access, because the concern was that tourism was beginning to arrive. The result in this context was a break, a violent rupture. The opening of the avenue, cutting through the medieval quarters. This was already, in itself, a disrupting feature, requiring something that had nothing to do with what I would do later at *Chiado* in Lisbon.

*Chiado* is a completely different case. The whole *Pombaline* downtown district of prefabricated buildings from the eighteenth century, featuring such great unity. *Chiado* is a completely different case, a convulsed, even tertiary site. The strategic planning decision was all about the tertiary sector, with support for mobility and tourism in mind. From the point of view of traffic, it was a key aspect in the Auzelle proposals; the plan was based on viaducts.

In fact, I had many problems with the first *Avenida da Ponte*, purely because we didn't incorporate viaducts. And that was what was preventing approval. At that time, everybody was going to Lisbon, to see the National Monuments. Because the building was obstructing the Cathedral, that was enough to complicate it. In the second *Avenida da Ponte*, the context of this intervention was completely different. What was intended was housing and the refurbishment of the historic centre.

They count for a lot, the personal decisions, the professional, the context you live in, beyond the historical moment.

**BR** – Let's turn to the second point, the methodology of the project. I take my previous question to be about something that seems obvious, at least for us: your path. It's a path of continuity, a path of adjusting to briefs, and a path of continuous learning. With some uneasiness, too, in looking for new solutions, new materials, new experiences, but in a somewhat austere, careful or even cautious way, maybe.

How do you see this?

**AS** – I believe that the methodological approach to architecture always has continuity, even though there are apparent ruptures. There are violent breaches that appear – and succeed – in historical moments that justify it. But behind these structures, new technologies, etc., there is always a line of continuity that will serve as the material for the times that will follow, and so forth. It's no wonder that you note that side of what I do, because it's a search based on continuity. Even if this continuity is fitful, full of intense novelty at times. Note, what is called modernism in architecture is imbued with an idea that everything must be new. For the heroic architects of the '20s, the '30s, everything had to be new, because it was a new city, a new spirit, for the new man.

Later, it had to do with moments like the euphoria and enthusiasm at the end of World War II. There's a whole mindset turned towards ‘everything is new.’ The clean sheet, the *Plan Voisin* of Le Corbusier, everything was new then. Even though now, with hindsight, we know that that was not what really happened. There is a kind of natural resistance. I can’t say inertia – that has a negative connotation – but there is persistence in the evolution of cities, in the evolution of architecture...

**BR** – There’s a sustained progression.

**AS** – Exactly. Not considering these moments of rupture, often caused by technological progress. I have never really courted this sudden change of language, of concept, driven by technological progress, which is very exciting, but I look at it with some measure. I’m quite conservative.

**BR** – Continuing to speak of your methodological approach and the role of drawing in this approach.

What does drawing represent in your manner of creating? What is your relationship with drawing as a tool in architectural creation?

**AS** – It’s exactly that: a tool. Also, I like it, it’s a pleasure to draw. There’s that too. But in terms of its usefulness for architecture, its professional use, it’s a communication tool unlike anything else. I don’t mean that that’s enough…far from it. There’s nothing else as quick, in which the creative connection between mind and hand is so direct. Often, you do things unconsciously that end up working... There’s a text by Alvar Aalto that I’ve quoted thousands of times. In this text, he says that, at certain times, he got blocked halfway through a project. He would then stop and do some drawing. And sometimes, from what he did in those moments, completely removed from the problem at hand, without even thinking about it, he would get back on track. Therefore, in my view, drawing is an indispensable tool in my work alongside others, the computer, the rigorous designs, the models. None of these tools is enough.

A scale model can be very deceiving. You have to do large scale models, small scale models, volumes, then large scale again, to experience the space. I almost always make a 1:1 scale model. So, I have the space in front of me.

This project is in Korea. To go to Korea is not easy. This is a museum that is now under construction. It was done with [Carlos] Castanheira. In the studio, they made an enormous model. And I would peep inside, sitting on one of those office chairs with castors.

**BR** – The Alvar Aalto quotation leads me to another question. Does everything in your work start with the drawing, or is the drawing, as in the case of Alvar Aalto, who would draw to relax, born from some kind of stalemate?

**AS** – All of this happens. It starts in the mind. Everything refers to the mind. There’s the use of various drawing tools that stimulate, that can open up unexpected paths … Basically, it’s the work of the mind. And as such I admit that it’s not strictly necessary to draw.

**BR** –You have a very close relationship between mind and hand, closer than normal…

**AS** – Anyone who draws a lot has this; it’s acquired.

**BR** – It is something you train for.

**AS** – Of course, it’s the same thing. This idea of ‘I have a knack for drawing, I don’t have a knack for drawing’ is nonsense. At the Faculty of Architecture in Porto (I guess it’s still like this, I’m a bit out of touch), you would do a lot of drawing. The drawing module was very important, and actually, the students, almost all of them, would end up drawing very well. Sometimes, in a cold environment, something similar could happen.

When this happens, something is wrong. Because the mind is in a different mood, the drawing is different. I admit that some architects don’t even need to draw. Mentally, or using other processes, they find a solution. The method of work is also very different from person to person.

**BR** – Concerning this direct mind/hand link … In various texts that you have written, there’s always the need to be drawing, to consider every eye, every view. Isn’t there a search for the material? Isn’t this an obsession for perfection?

**AS** – Sometimes it becomes an obsession, when things aren’t going well. There may even be an obsession, a certain anxiety. But there’s a point at which the project becomes difficult to control, even with scale models, which is when you’re trying to control the entire space. Not only windows, walls, there’s also the ceiling. And these days, the ceiling is a receptacle for a thousand things. How often is the air conditioning in the ceiling; the electricity, the lighting; you need to control a lot of things, so that you don’t end up with a space overrun by mechanisms. This is a lot of work and involves working with models, but the models don’t give you everything you need.

**BR** – Speaking of models, I’d like to introduce the question of sculpture, too. Where do Fine Arts and the search for form come in? This transition from design to form, when does it manifest? For example, at *Iberê,* which is a sculptural form, with all of those ramps. Seeing the building from the outside, you can deduce a thousand and one things, depending on your mood. How do you see this connection between design, sculpture and form?

**AS** – I see it as the result of architectural research into the space. The negative of the space is the volume, it’s form. So, I don’t associate it much with sculpture. In sculpture, the aim is different; it has nothing to do with function, the transition between spaces, it’s form. I think it’s a significant mistake to associate architecture with sculpture, except insofar as painting, sculpture, film, music, all belong to the same family as architecture – they are always communicating and always interacting. Take music, rhythm, the course of music.

There’s an architectural pathway in film. When someone is thinking about a building, they are also thinking about movements within the building: how people will move from the lounge to the kitchen, to take the simple example of a house.

These are the pathways that, in cinema, we see objectively controlled. Like the pace. They belong to the same family, but they aren't used directly, independently, in architecture.

**BR** – They arise unconsciously.

**AS** – They are all included in the architectural research, and I never thought, of this specific building or of any other building: 'this has to be a sculpture.' No, this form arose from several options, for particular reasons, from the desire to create a pathway external to the museum. A kind of pause in the visit to a museum, which, in different ways, is something that I value a lot. Museums are large, and therefore can be tiring. The very grasp of what's there requires an occasional break.

**BR** – But this structure of reasoning that you just mentioned, the idea of the approach to the works is very much Fine Arts. This route, this mastering of the form. Doesn't this have to do with your wish, when you were 18, to be a sculptor?

**AS** – Maybe, but I don't think so. I detached myself from any illusions about painting and sculpture. Among other things, because I was married to someone who drew and painted so well … It's not worth it, if there's someone at your side who draws much better, and paints much better, it's not interesting…

**BR** – I say this because – maybe I'm interpreting this incorrectly – but you have an enormous facility for solving practical problems in Architecture – for organizing programs; for addressing demands in regard to standards and regulations, for inserting structure, for defining those pathways first in the drawings and then in the models…

**AS** – I remember when my daughter was six, she had to do an essay on her father. It ended: 'My father makes architecture very easily.' It was the last sentence. For me, architecture is an art. No matter how much this idea is debated, for me, architecture is an art.

Therefore, this search that you refer to is not unrelated to the functional aspects. It's not, 'now this is sorted out, we can go on to that.' No. It's a whole. For me, architecture is an art. There are those who disagree.

Adolf Loos, who had a natural gift for art in architecture, said that architecture was not an art, for certain reasons. Surely reasons that were related to his opposition to the trends of architecture at the time, especially in Austria. He used this phrase, this idea. Le Corbusier had other ideas, which were then hard to see in some of his works.

**BR** – I recall Rietveld, for example. The reference to other arts in his work defines the architecture that he engaged in. Painting and its chromatic expression. In your work, it's more the sculpture, be it in regard to the form, the material, your background.

**AS** – De Stijl was even more than that. Rietveld would combine colours. He would use panels with colours, red, yellow, pure colours, and also horizontal and vertical lines. But this was a way of achieving an idea of what is, what has to be, architecture: Life in itself, the idea of anti-naturalism; of abstraction, the trend of finding the essential things… He has a series of pictures with a tree … He starts naturalistically, becomes more abstract, until the last tree is composed of vertical and horizontal lines. He was deeply focused on the quest for the essence, and for modern man, this essence was the machine.

It was necessary for art, for painting, for sculpture, for architecture. And he belonged to a movement, not made up of many people, but very strong in its intentions, in its aims and in its quest. It's interesting that Van Leuven, the architect, was part of the group. He had a lot of reservations about this association between painting, sculpture and architecture. The same idea behind the transformation of man struggling so as not to surrender to nature, of capturing the essence of what nature means.

Van Leuven thought that architecture was a little different.

**BR** – In the *Iberê Camargo* video, there is a phrase that goes like this: 'Siza, the master of the masters.' I'd like to take this phrase and move on to Siza the master builder. Let's imagine that you were a man of the seventeenth century: what would it be like to be in charge of a piece of seventeenth century construction work? Do you feel like a master builder, in the seventeenth century sense?

**AS** – In the seventeenth century, they probably didn't work with engineers. I don't know… In the seventeenth century, they wouldn't have this division of work: that's the architect, that's the engineer. There was no such division of labour to the point that there is today, between those who build and those who design. The architect had a much more peaceful life, because it was enough to make a sketch and give a verbal explanation, and the master-builders, the constructors, the labourers would develop it. The exhaustive detail that's required today to convey the idea wouldn't have been a part of it, the fragmentation that today makes up the team, so large, even universal, that it's not enough to just talk. It's the coordination. It's the heart of the work, today. I remember when I started working, I worked with constructors, master-builders; there was almost no way the architect would arrive on the site, see something badly done and say 'take it down.'

Because, if the master-builder saw something badly done, he would take it down. It was shameful, a tremendous loss of prestige for him if the architect had to say something was badly done. I've worked in close contact with workers, with labourers, with whom I learned an awful lot at the start of my career as an architect. When I worked in Berlin, in the 1980s, I went to the site, and as was my habit, I started to talk to a tile layer. He said, 'In Berlin it's not done like that ...' He ended up thinking that, actually, it was more practical, easier, to do it my way. As I was saying…I got to the office and got a call from the Site Manager, the Contractor, furious, saying 'You can't talk to the workers…'

It was then that I understood the change. A pecking order was evolving, the division of labour. Everything thought out in terms of strong coordination, a coordination that, for me, has to be part of the actual design.

**BR** – But that makes it hard to work. Do you feel like a master of masters or not? In all ways except in the contact with the workers?

**AS** – I think that concept is absurd. There is no master of masters. No one can say who the greatest architect is today. Never could, probably. Was it Frank Lloyd Wright or Le Corbusier, all those years ago? You can't say that.

Today, with the multiplication of activities, experiences… You can't say. It's impossible. Or it's wrong. I love the way it all works, regardless of occasional incidents, but I never had those with a design team, or I don't remember having had them. I work with the same designers, but I vary them sometimes. There really is a great understanding from the start, and the architect is the coordinator. One thing, though, that is more worrying for me is that with this recent review of our professional status, the word 'coordinator' has disappeared. The idea of the coordinating architect is not there, there's no accountability. So who is the coordinator?

**BR** – That coordination requires strong intervention during the construction phase. Do you agree?

**AS** – And there can be more and more constraints. I've been looking at a notice from the Architects' Professional Association, about working with an architect. It says there that the architect, as the designer, cannot be the supervisor. This is simply the enshrining of an idea that is current practice. Today, I could visit a site where the supervising team would say: 'No, no, this is our job.' A monstrosity. It's a harsh blow to the possibility of dealing with this.

**BR** – Going back to the question of contact…

**AS** – Of course, you can also think that this business of building in a certain way, in our environment, is a thing of the past, an old-fashioned idea. You can think houses are cars. I don't.

**BR** – Basically, to take tradition as a challenge for innovation. How do you see this change in architecture?

In the beginning, you had time to talk with the master-builders, the metal worker, but you don't have it any longer. How do you manage this dialogue?

**AS** – I have less time because I have more work. I have other resources, a team. But that's not all. Above all, it's just no longer possible. It's not allowed. It's not like it was in the time of the craftsman on site, certainly not. But then, to completely cut the link between the thinking and the doing, that can't be allowed either. It's done by other means, naturally, working with industry, for example.

**BR** – That's why, in many jobs, you need to take the design to the end. To design everything, from the walls to the furniture, the cutlery, the plates …

**AS** – In the Corbusier sense, *'la machine à habiter.'* When you make a machine, when you make a car or a plane, everything is thought out down to the last screw, to make it work, and work well. The idea that drawing for us is a thing of the past, or an exaggeration, this is a lie.

**BR** – I'm talking about design. I remember that you designed the plates for the Portugal pavilion at EXPO98, for example. Father Higino, from the *Church at Marco de Canaveses*, asked you to design his clothes ….

**AS** – Not necessarily, it doesn't always happen, only when people have particular aims in mind. I only do it when people ask me. I don't impose the furniture upon people. Terrible things also happen in this field. You're doing a job, a hotel, for example. At one point, they say, 'There's already a team of interior decorators for the interior, this has to be comfortable.' Then, a team of decorators (it could be a good team, I'm just talking generally) enters into this new arrangement of participants in the architecture, in the space.

They say the exterior is for landscape designers, the inside is for interior designers – I'm not saying decorators… And, as someone said, I don't know who, we are left with the skin! There's no doubt that the interior is born from the outside and the exterior is born from the inside…

It's the end of architecture, pure and simple.

**BR** – There's no space for architecture.

**AS** – The impossibility of architecture … it's not stealing the space…

**BR** – Even more serious.

**AS** – Of course, you work with the landscape designer, with furniture, decoration specialists, etc. You, do this! You, do that! I began to understand the attacks on the work on *Avenida dos Aliados*, because one of the slogans was: 'Give the exterior spaces to the landscape designers.' When I saw 'this thing' that entered Parliament, I understood!

## Architecture and Engineering

**BR** – That gives us a bridge to the next topic, architecture, engineering and other players in the work. Let's start with architecture. How do you work with your colleagues? What kind of links do you have with them?

**AS** – It varies. I've had some colleagues for 20 years. More than 20 years. There are others who've been here 3 years. How long have you been here, Bárbara?

**BR** – 5 years.

**AS** – That's not bad! Normally, there's one, or more than one, depending on the complexity of the work, responsible for conducting a particular project in the office. When it's necessary to strengthen that, others join the team, generally younger, less experienced. I keep an eye on it, I watch over everything. It varies over the course of each project. There are projects that don't take up much energy at a particular time, but then there comes a time when they need particular assistance.

Therefore, it's not possible, I would say, to plan. There are projects for which it's completely impossible. There are projects that stop, and then re-start. You have to deal with it almost by instinct.

**BR** – Flexible management. And the question of other architects working on a clear co-authorship?

**AS** – In this moment, I have four jobs: António Madureira, with whom I did the Bouça, Carlos Castanheira, with whom I am doing the jobs in Korea and Japan, and Souto Moura, in the Avenida dos Aliados, for example.

**BR** – But are they co-authors?

How does this co-authorship work, how does the creative process function? What about planning the work?

**AS** – The results work very well. These are people who don't have any egoistic conflict about authorship. They don't raise this sort of problem. Let's speak about the results: I did the *Serpentine in London* with Souto Moura. People told Eduardo, 'This doesn't seem like your work.' And they said the same to me, 'This doesn't seem like your work.' Of course not. If we wanted it to seem like my job or his job, we would not have chosen to work together. It is obvious! It's very stimulating, very exciting, to get out of a certain routine, the day-to-day routine of the office .... To escape from the office, from that routine. It lets you really get away from the routine.

**BR** – Let's move on to the Engineering team. This is commonplace, but I think that you like working with stable teams of engineers. Because they've become friends, because you like working with them right from the start?

**AS** – I like working with good engineers, good people. And that's what I've found. I normally work with GOP. We're associates; they're here in the same building. I liked working, the first time for me, with Prof. Adão da Fonseca, on the bridges. He's extraordinarily nice, we got on very well. I think he also liked working with me. I don't think, I'm sure. He liked it, and even wrote as such. Why? Because there really weren't any, how do you say it, biases. I could drop an idea and say, 'Is that nonsense?' Because a bridge is not a small structure... We worked well, on the basis of hypotheses, checks. It was an excellent experience. A job that I liked a lot.

**BR** – It's like playing with four hands ...

**AS** – And just as well, because both hands, of course are Adão da Fonseca's! I say this, in relation to my participation in this work, precisely because it was a... an uninhibited participation....

**BR** – Moving on, now, to the question of regulations and requirements. Are they a prison for you, a release, a reminder that you use to be daring?

**AS** – For me, all constraints are good project material. The limitations of the terrain in the Iberê Camargo Foundation project were difficult. It was this difficulty that largely led the project. The constraints are a foothold, a basis for rationality. But lately, the issuing of regulations has reached a fever pitch. Every new day, a regulation is published that cancels out another. This, I think, is debauchery. I did a project for some holiday houses, small apartments, in *Parque de Vidago*. We did 4 designs. One was finished, with direct consultations with the authorities, then a regulation was established that made it impossible. In some cases, even with planning permission approved. Some of the rules are absolutely preposterous. It seems like it's almost a compulsion to introduce regulations in torrents, hastily, without reflection. And then change them.

Everything goes back to the drawing board, or to a more convenient board. This is disturbing.

**BR** – But often, it's also stimulating.

**AS** – Yes. All the constraints are stimulating, and are material for design. But this unruliness, it's not good.

**BR** – In the FAUP's project, by the end, one example of technical conditioning became an added-value to the design: the sun protection of the facades....

**AS** – I don't remember this.

**BR** – It was Eduardo who told me. The Vice Chancellor's office passed the design to FEUP for technical evaluation.

**AS** – At that time, there wasn't the same burden of regulations we have today. What I blame today is the stampede of current regulations. The country was terribly behind from this point of view on issues such as insulation, energy, etc. Nor could it consider these issues in general, because it required immediate investment. Therefore, almost no one wanted it, especially in council housing. I remember that, in one of the projects I did for the *Alentejo*, not *Évora*, it was *Vila Viçosa*, the design was ready, and then a regulation was established saying that all houses had to be suitable for disabled people. I had to redo the whole project. Of course, that was before the elections. I created a new design. Then, after the elections, the regulation expired, and I had to revert to the first design, with small alterations.

**BR** – Still talking about the team. How important is the developer, the client, in a work of architecture?

**AS** – Without a good developer, there's no good architecture.

**BR** – And is a good developer, basically, one whom you trust and who gives you room? Or is it one who's very involved with the work?

**AS** – It's the one who is very involved and has no biases, who has the desire to build well, to do a good job, not only within the regulations, but with a real sense of quality. I can't think of a better example than the *Iberê Camargo Foundation*. People really wanted to make that museum, and had been friends with the artist; they wanted to do a good, capable job. The builders immediately said they would form a team comprising the best carpenters, the best stonemasons, etc., etc. I knew there how far we could go in terms of demands. I had a great deal of support, even concerning relations with the City Council, approvals, etc. They negotiated things … It was an exceptional case … I don't imagine another one like that will ever come my way.

**BR** – Regarding the cost of the work. Is it a false problem? The developer is the one who must define the brief, he has to know what he wants. At the Foundation, they knew what they wanted and that the cost would result from what they wanted. You can't want something expensive at a low cost.

**AS** – How can you do it for 500 € or for 550 €? You can't. You know you can't from the outset. Sometimes, there are subtleties to establishing a price that's tied to the contract, and therefore we must accept that the work costs so much for …

For me it can be an illusion, but I think it is not. People can learn, because they may not know and they should talk to someone who does know. Other times, it's a way of offloading expenditure overruns on the architect. This is the case when the brief is suddenly changed, adding more area or more expensive elements. Then, when it blows up, the overruns … it's the architect's fault!

**BR** – I hope this has nothing to do with what you spoke of earlier, time running away. In Portugal, projects always seem to be yesterday's projects, and you have to do everything immediately.

**AS** – …Works are expensive. If things are not fully defined from the start, it's the job of the contractor, almost always with good reason (I don't mean almost always, but always with at least a degree of reason) to demand extras. Then, it's the fault of the architect, the engineer, or I don't know who in the design team! You're not given time to do a project well. This is a general problem, very general. Even with very large efforts. Some time ago, there was a notorious case of a competition that was open for only 9 days.

Actually, the Architects' Association has taken a position on that. It should have been the developer who was called upon to explain this situation, in this case, a City Council. This didn't happen. It just opens the way for cheating, with many possible ways of cheating.

**BR** – To conclude this topic, one final question: how do you see the architect in 2025?

**AS** – Probably, in 2025, I won't see anything. I'm no Manoel de Oliveira or Niemeyer. I don't know. There are major transformations underway. I think, first, that it won't be universal. Although it's fairly universal, there are nuances; what happens is not always like we were describing. Within this idea of continuity in innovation, I don't see the architect as being different from that. Basically, basically. This thing I was telling you about in Berlin, it could just as easily have happened in the Netherlands. I worked there on a council housing project. It was all prefabricated, all the building elements were prefabricated. But let's see: this was also what the Marquis of Pombal did in Lisbon, in the eighteenth century.

In the Netherlands, I had to choose the materials. I'd choose in the street, 'the brick, I want this, I want those windows, that colour.' I was used to going with the painter to fine tune the colour. That's over. There's a lot that's going to change, and especially today, with all the great intensity and the new materials. Basically, what happens is that the man is the same, or at least, almost the same.

**BR** – It's a new learning curve. Yesterday, while preparing for the interview, we came to the conclusion that this multidisciplinary team is the master-builder after all.

**AS** – When it goes well, yes.

**BR** – In principle, it goes well …

**AS** – You're an optimist…

**BR** – How do you regard this lack of time?

Not being able to control the timeframe like you did at the start? How do you cope with this lack of time now?

**AS** – For example, in a competition that takes place over 9 days, I don't, I can't. Lately, a lot of work has appeared: calls for tender, particularly from abroad. I have to say 'no.' It hurts to say no, but I have to do it. There's a limited amount of work that can be controlled. Depending on the choices you make, the type of practice you need, the team that you want. I have 25 people … already hard to control... It felt good when I was working with 10. But times are different. Now, there's a limit. With so many offers, one has to say no. It's impossible. Besides, it's more logical, the work has to be shared out.

Some have a great deal of work, others don't have any work. It doesn't make sense.

**BR** –You spoke, with a certain nostalgia, of a time when it was easy, allowing close contact with the workers involved in the building. Now, that's forbidden, or at least difficult. How positive are different forms of contract, like design-and-build or public-private partnerships? Or, on the other hand, do they give absolute power to the client, so that he can demand things that might be unaffordable?

**AS** – I never did … actually, I lie. I did. And it didn't even go very badly. It's very slippery ground, because at some point, it could lead to designers who are too closely connected to a contractor, so that you would hear very often, 'no to this,' 'no to that…' It's not exactly like that, because the builder himself is also, eventually, interested in the quality of the work, so as to win over the tender. It's a very delicate field. I think that it's early days, and has thus far shown very poor results. Another thing that has nothing directly to do with this is what's increasingly happening in competitions. To go for the cheapest is commonplace. It happens all the time.

This link between the builder and the architect may be fruitful. Extending it to the industries is a field to be developed, and it already is developing to some extent. Not so much in our country, but then, we don't have much industry, we import a lot.

This is a different way to restore integrity to a project, a building, by means other than the old way of direct contact with people.

You may miss the emotional side of things, but not the rational side. Not today. There are changes that can't be overturned. It's a tough field. Particularly in our sort of activity. The fee tables are done with, they are no more. So, that is happening a lot: whoever will take 1% gets the work. Of course, for a job to be serious ...you need fair payment.

## Museums

**BR** – Shall we move on to the idea of museums? The idea of museums was we were going to sweep through the museums, a little like examples of …

**AS** – Sweep … into the trash?

**BR** – No. Analysing them one by one, as examples of different briefs, always within a logic of continuous learning.

Let's start with *Santiago* and the logic of adapting the brief to the old city.

So that we understand how we arrived at *Iberê*.

**AS** – *Santiago* was to be a museum within a garden. The land was part of a convent that's a national monument of Spain. However, the garden was in ruins. The idea was to put a museum inside it. This is a common reaction, the fear of architecture: to hide it. I had the view that a museum is a building of the city, feeding on this affinity, so it shouldn't be hidden. So, I struggled to bring it forward, to pull it to the fore. In that case, it was what seemed to me right. But *Serralves* was different. In *Santiago*, it was clearly right to bring the building forward. Put it in a parallel position with the convent, a national monument, organise them as a whole. I was lucky that they also gave me the garden, and so it was a huge area, like a whole, interactive, in its transformation and preservation.

The importance of water, and the existing water channels there, helped a lot. The way the water is distributed reflects the whole topography, which helps in the interpretation. And indeed, the organisation of the garden, which is independent and features paths that kind of zigzag, basically became the basis of the organisation of the museum itself. The museum has some sinuous accesses that command a view over the city from the terrace.

It was very challenging to work side by side with the architecture, in that case, from the seventeenth century, but in the centre, completely surrounded by mediaeval buildings. It had an influence on everything: in the material used, in that case, the stone from Santiago. In relation to the organisation of the museum, I didn't have, from the outset, the possibility of exchanging ideas with the director of the museum. Actually, there wasn't one, he only appeared later. So everything was thought through on the basis of the study, the museum theme, that not very old theme.

The problems of light, designing for a great variety of uses, the collection of the museum, facilities, more classical exhibitions…A space that had a particular form, rooms, generally rooms with different articulations and different sizes, a clear route, so as to avoid having signs everywhere. From the point of view of movement, there was a path through the museum, and also partial routes. All the flexibility that is required in a museum, and all through study, but no dialogue with a director.

In *Serralves*, it was similar. There was a committee that provided guidelines, but there was no museum director. It was, therefore, also very personal, depending on the data that was provided. In this case, the siting was very important – it had to respond to various points of view. To have a side entrance. I wanted to make the entrance in the centre, at *Gomes da Costa Avenue.* The compromise was to make two possible entrances from the front courtyard, to be used according to circumstances. The organisation of the interior and the light had a lot to do with the experience in *Santiago.*

But again, there was no one with whom to discuss it. This happened in *Stedelijk* with an extraordinary person I met (indeed, a consultant who was called, at one point, by *Serralves*), Rudi Fuchs, an experienced man.

Overall, it was the most gratifying and stimulating experience. At least the solution for the museum was discussed step by step with the curator, as far as the museum's idea for its project was concerned. It was very interesting.

But, nevertheless, there was a political change in the City, and he left. Then, the architect left …

**BR** – That explains it. The most difficult developer, the political one.

That takes us back to *Serralves* again. You did five designs for *Serralves*...

**AS** – Yes, there must have been five Ministers of Culture… No, not that many, but at least three.

It should be larger, it should be smaller. It should have a conference centre, it shouldn't. It was very hard …

**BR** – And without a fundamental interlocutor, the curator. One of the basics.

**AS** – One of the basics. However, I must say this: in that respect, it's fairly common, and brings discredit to both parties.

The example of designing a kitchen for a chef. A kitchen is designed for one person, who determines all of the requirements (I remember this at *Boa Nova*, all those years ago!) In the end, he didn't stay in the restaurant, so then someone else came, and it was all wrong! Curators will also change. In *Serralves*, there was more than one: there were two, and with a certain degree of continuity between them, in this case, but that doesn't mean that there won't be another with different ideas. I think both *Santiago* and *Serralves* demonstrated their openness to more unique types of exhibition and resilience regarding maintaining a coherent whole. Whatever the interventions they may engage in, often closing windows, switching off lights. They have a certain consistency, I think, both of them. They have worked in many different ways.

**BR** – Don't you think the absence of a client, the absence of a developer, is complicated, in the design of a contemporary art museum, compared with a classical museum?

**AS** – No, I don't think so. For example, it wasn't in The Netherlands.

**BR** – But the curator was present from the beginning…

## Maintenance and Operation of Museums

**AS** – There are periods of transition, as we mentioned earlier, when the design of the museum is not working. Then, there's an adjustment. In this case, the first reaction is strident, because the trend is to change things… The person has another project in mind other than that, and thinks that the result will not match their wishes. The problem with the museum is that it has to have good conditions of light, good routes, distribution of galleries, flexibility of use. But this only holds if there's consistency. Some people think (not anymore, I think, but it used to be like that) that the museum should be a large space in which each exhibition is subdivided and there's total freedom. This has been tried and tested, I think, and it has been concluded that it doesn't work. It's what justifies the *Beaubourg*. That was done in that way, a huge space. But it has been refurbished twice, both times in order to reduce the dimension of the spaces. Also because of maintenance cost issues.

**BR** – I.e., the architecture is relatively stable, and what is on offer at the museum is very variable and very adaptable to people and rules.

**AS** – In a museum, the space has to be able to receive very different things... In the *Iberê Camargo museum*, they did – in fact, I really like this building, it's a sculpture. They used the central space very well. I'm very satisfied. This demonstrates how a museum adapts itself to the concept of the building. Some people said it was very formal, that it was restrictive. But there have already been three exhibitions, and the feedback I get from there is that it all worked very well.

**BR** – In the end, the museums came back to a classic concept. After many tries, the concept of a contemporary art museum came back to a classical museum. I'm thinking, for example, of the Iberê Camargo Foundation – which you have given a badge of a "hidden classic of architecture." The Tate, by Herzog, is also a very classic museum: it has a very large atrium, a very defined route through the galleries and the architecture and the art function very well.

**AS** – It varies a lot. In cases when there was a director, a curator, things went well.

The difficulty here is that the curators change, and therefore there must be the same flexibility, because another comes along, who then has his own demands. There's always a need for a certain distance in relation to the role of a museum project. There must be a consistency, a mark in the design; otherwise, it will not stand the passage of people. But there are many ways of looking at the relationship, for example, the nature of the museum. There is a beautiful museum in Denmark in the middle of a garden, all glazed. The curator of this museum was consulted by *Teresa Patrício Gouveia* (CdO – The Minister of Culture at the time…) at the start of the *Serralves* project to do the first brief. Then, it was abandoned, because it was a little museum. Then, there were all those changes... The Renzo Piano museum, in Basel, has a huge amount of glass. Usually, there's resistance to it, regardless of whether there's a curator, there's resistance to it. While I was doing every one of these museums, they would call my attention to two things: one was that there must be walls, the other was the windows. There are other aspects – and the other aspect was 'be careful with the architecture, don't try to vie with anybody...'

**BR** – To outdo…

**AS** – Exactly. This leads to difficulties sometimes, because if the museum is pleasant and if there are visitors who are coming to see the architecture, and possibly the exhibition is not so great, then an interpretation arises: it's the fault of the architecture. But when there's a great exhibition, no one is distracted by the architecture, because they are going to see a good exhibition. Moreover, it's something very strange, because no one reacts to the strong character of many museums – very strong! In the Louvre, no one says that the architecture is stifling the work of art. This is a bias, an invention. Although it can happen: the architectural design can be so strong that it is excessive; that's the fear of the design. The design is not one window more, one window less.

**BR** – How did you deal with experiences like *Santiago* and *Serralves*, which didn't have a personality behind them? Painters like Cargaleiro, Nadir Afonso, Iberê Camargo: They had a very strong personality behind them. Is there any contamination of the artist in a museum dedicated to a specific artist?

**AS** – Oh, yes, the artist himself? I can't always say that there isn't. Consciously, voluntarily, there isn't. Unconscious, now, maybe. I saw the work of Iberê Camargo, of Cargaleiro. This pervades our thinking a little, but not as an objective.

A museum of contemporary art, even one that's primarily intended for the work of one artist, is not just limited to putting his work there. Otherwise it flounders, it falls.

**BR** – But the *Cargaleiro Foundation* had to be for *Cargaleiro*. The form of the building is so organic and complex…Nadir Afonso's foundation building is better understood. Maybe one understands the relation between Cargaleiro and his work and Nadir Afonso's building with its more sober character…

**AS** – They already told me this, in relation to *Nadir Afonso's museum,* but this was never my purpose. Of course, I know Nadir Afonso, I know his painting, and there is some influence.

**BR** – I want to ask a question, a topic that is particularly dear to me, the difficulty of adding to a brief in a classical building (we are back to the Netherlands) with modern architecture. That, to me, not being an architect, is the most difficult thing. I'd like to hear what you have to say about this.

**AS** – The thing is that they are all difficult. To do *x* house in *y* street is a problem. But doing a house in the desert is a problem, too. They're all difficult. That is, they require concentration. Speaking of the museum in the Netherlands: enlarging the old building was also a form of stimulus. It's another conditioning factor. For those who want to face it, it's a stimulus for the project. It's like the landscape, the case of the hole in *Iberê Camargo*. The *Stedelijk* is a very eclectic building, very elaborate, very clear. I made connections to the extension at two points, almost like bridges that entered into spans in the building. Two spans, one of them a very large window. The bridges were there exactly so as not to touch the architecture. It's clearly in contrast, but it's a contrast that is not an abuse. Rather, it's a contrast that blends like a connection that is not ...

**BR** – It's an association that does no damage.

**AS** – Yes, a connection… Forming a whole, but forming a whole without copying the language of the architecture and the materials. It's more a relationship of contrast, which does not result in fragmentation. On the contrary, it results in a whole.

## The Ibere Camargo Foundation Project

**BR** – Let's move on to Iberê Camargo. I was lucky enough to work with you from the start on this project, and I remember that, in the beginning, it was in a very different form. The structure, the organisation, a box that was being improved and adapted to the site, to the place with a mountain behind it. I noticed that, many times, you worked with an almost three-dimensional chart that was gradually contaminated by various things, whether it be the relationships of the architecture that you grasp for that specific case, or the terrain, or the landscape. What happened in this case?

**AS** – What happens in most projects. You feel whether the project has density. A site, the context of its formation, the kind of brief, etc. When you start a project, you don't know what you will do. You won't do something like the previous one. This may happen eventually, but usually not. What I realised about the vision at the design stage was mostly through contact with the students, through the problems of how I ran the classes. If someone exhausts their knowledge of the problem before engaging in the project, before having any idea of how the building will turn out, all that information is channelled towards a solution, i.e., it stops the research almost immediately.

What I try to do is not to rule out hypotheses, no matter how oppositional they may seem with each other, at a stage when I can have the freedom to not yet know the brief well, the problems of the project. I can thus play around with different ideas for solutions.

Each one, as it comes to mind, needs immediate testing, and therefore requires more information. Adding information then leads to discarding this, discarding that, etc. I try to give myself as much time as possible to develop a very wide range of ideas, which then go through a critical process that will allow for further information, for the chance to go truly deeper. Until we get to a point when we believe we have the necessary information at hand and ...

**BR** – And those possibilities strengthen the one you want to have?

**AS** – No, no, no. I don't want that. It's gradually defined. For example – it is related increasingly to function. One of the first ideas I had was for the car park to be on top. You'd go down in the elevator, entering it at street level.

Later, we found that it was impossible to do that, because the land there wasn't affordable and there would have been a reaction ... It's a wealthy residential area. So, that idea was scrapped. But what was done, you can see that the sketches of this idea didn't disappear entirely in the other phases. There are drawings in which you see a group of different solutions on the same sheet of paper.

**BR** – The ramps were always a constant. But at what point did you let yourself be contaminated by Brazilian architecture, especially by the rules of Lina Bo Bardi?

**AS** – This is funny, because we talked about it and I said 'right!'. I had visited her building, but the Lina Bo Bardi design never crossed my mind. It has inevitable similarities, because first, it's the building that stands out, and second, because, for its own structural reasons, it can't have large spans, so it has some little holes. What I said, when my attention was drawn to the influence of the *Lina Bo Bardi building*, was this: 'No, no, we were both influenced by the *Factory Van Nelle building* in Rotterdam. It was the first time I saw (I saw by myself, not with Lina Bo Bardi) two buildings connected by bridges like this. Clearly, an architect's training, something that's never over, comes from seeing a lot of different things. You see so much, on so many pages, that you don't even remember it, but it's here, at the back of one's mind, it's not asleep and will appear sooner or later.

**BR** – The form of the curve in *Iberê Camargo* clearly comes from Brazilian architecture…

**AS** – No, it doesn't. This form comes… A desire for curves, yes, maybe. We went there, to Rio de Janeiro, to see Niemeyer, etc. But if you look…this curve, in the front, is the negative, it's the symmetrical counterpart of the curve of the mountain. At a certain point, when the curve was fixed, as you well remember, that was hard work! This is the curve that is right here, making its way inside in a dialogue with the curve, it's then that… I want to show you something here… Guggenheim, Frank Lloyd Wright… I didn't know this. It was a trainee from Brazil who brought this to me.

He started with a hexagon. Basically, in this case, it's exactly the same thing. I was just looking for the shape. This here had to become a circle. It's exactly the same organization.

**BR** – When we were working, you often spoke about this Guggenheim building, the ramps…

**AS** – There was something fundamentally different, which was … and the ramp was separate.

**BR** – To bring everything together in the outside form.

**AS** – It was highly criticised, it wouldn't work.

He, Frank Lloyd Wright, didn't go to the inauguration of this building; they had fallen out.

They changed the lighting completely. The light would enter from below, the light would be direct…

**BR** – Direct and natural.

**AS** – They didn't allow it. And so he didn't attend…

The form is the same, but inside, the light doesn't come from there and it has terrible lighting.

**BR** – The question of white concrete: did you like the experience?

**AS** – The white concrete is a colossal job by both the engineer here, Jorge Silva, and the one there, who was a marvellous guy, who understood us very well. We got along fine. I'd done some swimming pools in Barcelona in white concrete with Jorge Silva. He went there many times to fine tune slabs, materials; and he managed it. It wasn't bad in Barcelona, but it's better in Brazil. In Brazil, white concrete had never been done before. The climate is different, so they conducted some experiments, including using parts of the car park that were meant to be covered to do experiments – experiments that were intended to stay, that helped fine tune slabs, formwork, other problems.

**BR** – They were very receptive to it, we imagine…

**AS** – They were very receptive to it. They took the shuttering off and put up some dampened sheets, because of temperature differences, among other things. It was research done locally. In Barcelona, Jorge Silva played a key role.

**BR** – And there, you worked with the Faculty of Engineering in Porto Alegre.

Speaking of comfort engineering, mainly thermal, how was this project designed technically, considering that it would be working in a completely opposite situation in terms of thermal behaviour?

To avoid the heat, look for the cold, the problems were inverted…

**AS** – The engineer there took care of it. In fact, we also did a house in Majorca with the same problems. I didn't want to put air conditioning in, but it's a kind of law. I think they never turned it on. It's a linear house, long, it couldn't be done any other way, because that's the shape of the land.

**BR** –How do you see this issue of sustainability of the work, the project?

**AS** – I'm fine with it, but even if I weren't, it's required.

**BR** – Do you see sustainability as a duty of the architect? Or was it already a duty of the architect to have a constructive consciousness?

**AS** – I think so, because in any case, a construction… Of course it is. For a long time, there was no problem, like there wasn't with cigarettes; they were bad for you and you still smoked. With comfort, it was air conditioning, cartloads of air conditioning and the problem was solved. You'd go out into the street, like in Brazil, and you'd catch a cold, guaranteed!

You go into a foyer in a hotel in Brazil, and it's like ice. They're adapted to it there, now. It was the idea that differed. The idea was that there are technical means to achieve comfort; whatever the solution.

**BR** – My question… That is, sustainability, as it is understood now, involves a series of principles that may already be obvious in architecture of minimum quality.

**AS** – Yes. It works, the architectural solution…

**BR** – But are they internalised in the way you work or do you talk to Jorge Silva, to call attention...

**AS** – I'm the one calling attention from the beginning. The plant room, the size needed, studying the routes, avoiding grids everywhere, studying the routes...Ceilings for passing pipes, the height of the building, from the start…

**BR** – I remember, for example, there were reflective walls in *Iberê*, and I remember that in that pavilion, you had to abandon the idea, of APDL, which was to use river water to....

**AS** – But this was put there, they are impositions of the engineer. I don't know if I have time to make art... Like my staff say, I have time to think globally because I have good engineers. I never thought of using the water ... You hear about it and know it happens, but I don't worry about it. The engineers work on the project from the beginning, like with the lighting. In museums, it's extremely important. I have a capable team, good, very good. And I don't worry, because they warn me to be 'careful because'....

**BR** – Therefore, it's important…

**AS** – It's fundamental. The design today can't be done by the architect and then just passed on to the engineers.

**BR** – I have a question I want to ask, I think it'll be my last one. It has to do with the client. One of the things I liked seeing in the video there in the *Iberê Camargo Foundation* was a lunch they held with everyone, with all the workers, to celebrate… I'd like you to talk a little bit about the spirit of that job, from the point of view of the way in which the client worked, the relationship with the workers, how everything was fantastic. The idea that we have...

**AS** – The main sponsor of the museum is a very strong industrialist in a company that recycles steel, GERDAU. It has plants in Canada, Brazil, Argentina. It's a hugely successful industry. One of the reasons for the success that I saw there is that the relationships in the factory in Porto Alegre are extraordinary, it is like a family. Labour relations are fantastic. He's a guy with great vision. He was also an Olympic equestrian champion and has a wonderful stud farm ... and a great concern of his, indeed, it could be seen from the studies he did, to do with the museum, is having an extraordinary team relationship. And it also happened in the work. The relationships among the site manager, this engineer Canal, and the whole team, the workers, were fantastic. He had, at the same time, authority and firmness, but a great relationship, which is very much the Brazilian spirit. When a certain phase was over, he'd throw a barbecue, bringing together all of the staff, he'd take photographs of the whole team. It was an extraordinary experience. And there was no management team saying that the architect was late or that it was expensive, and other such things...

Without a good client, you can't have good architecture. It's impossible. It's a struggle that has to be reflected in the final result.

## General Arrangment Drawings

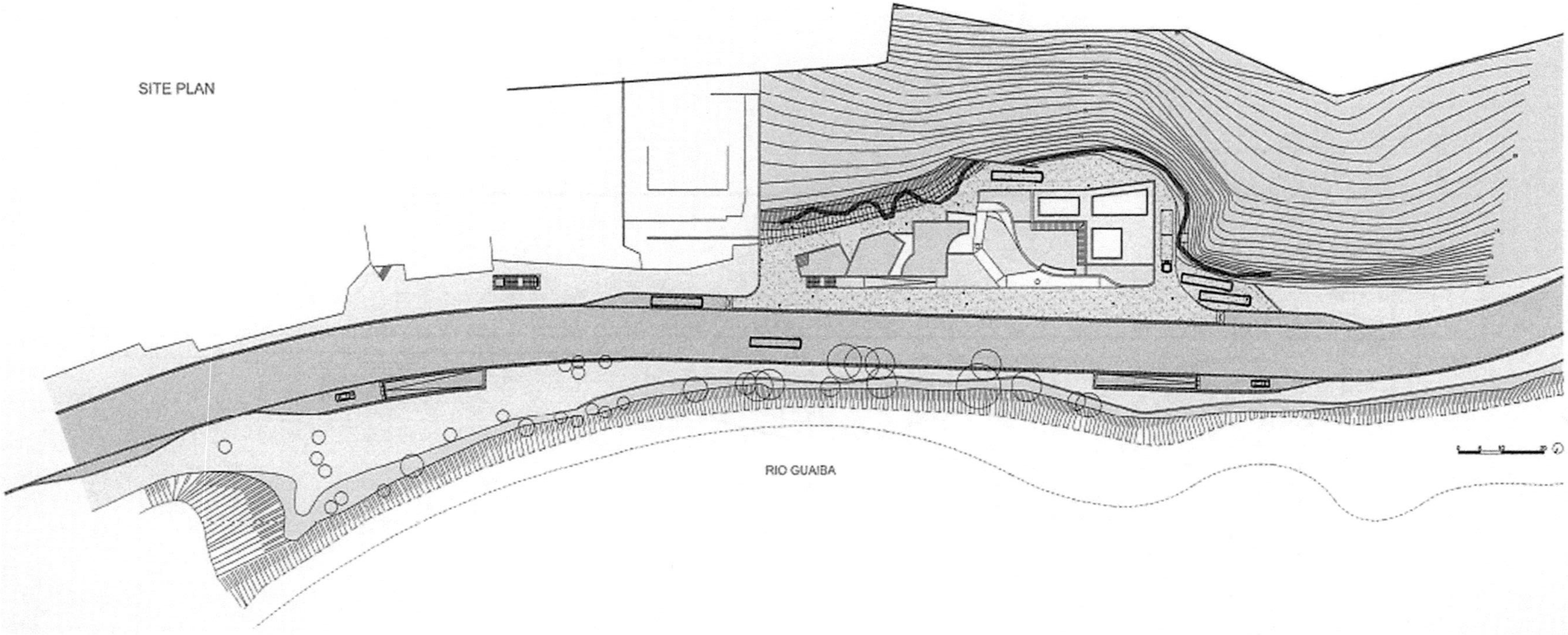


Site plan

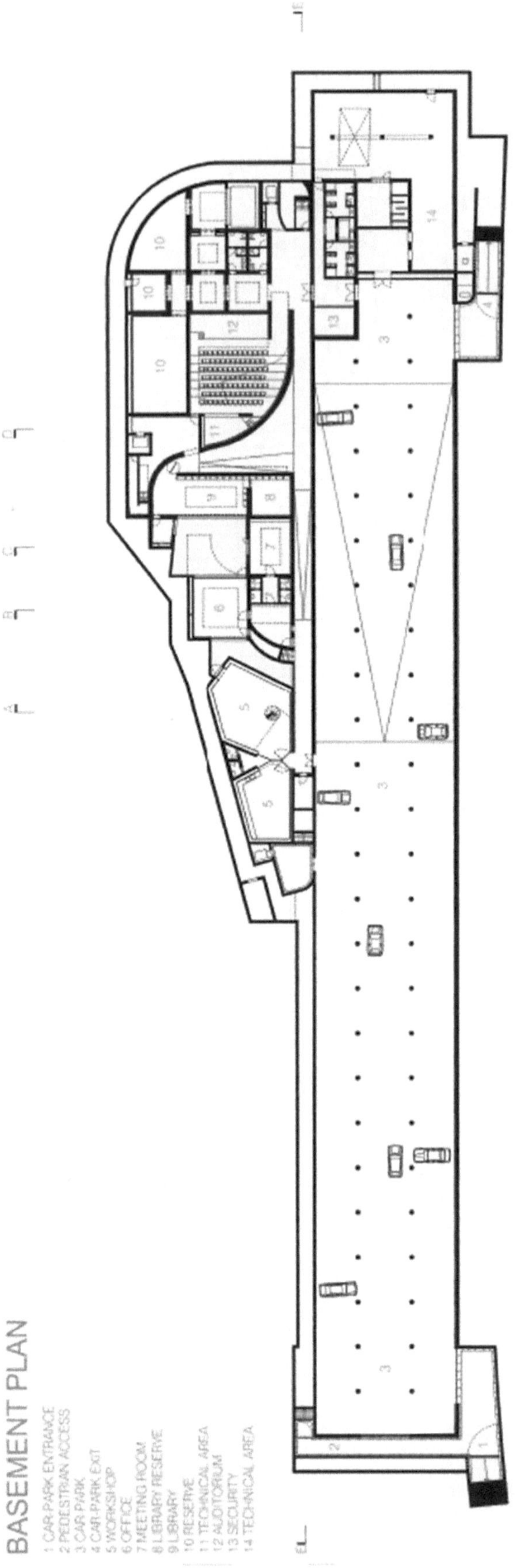


Second basement plan

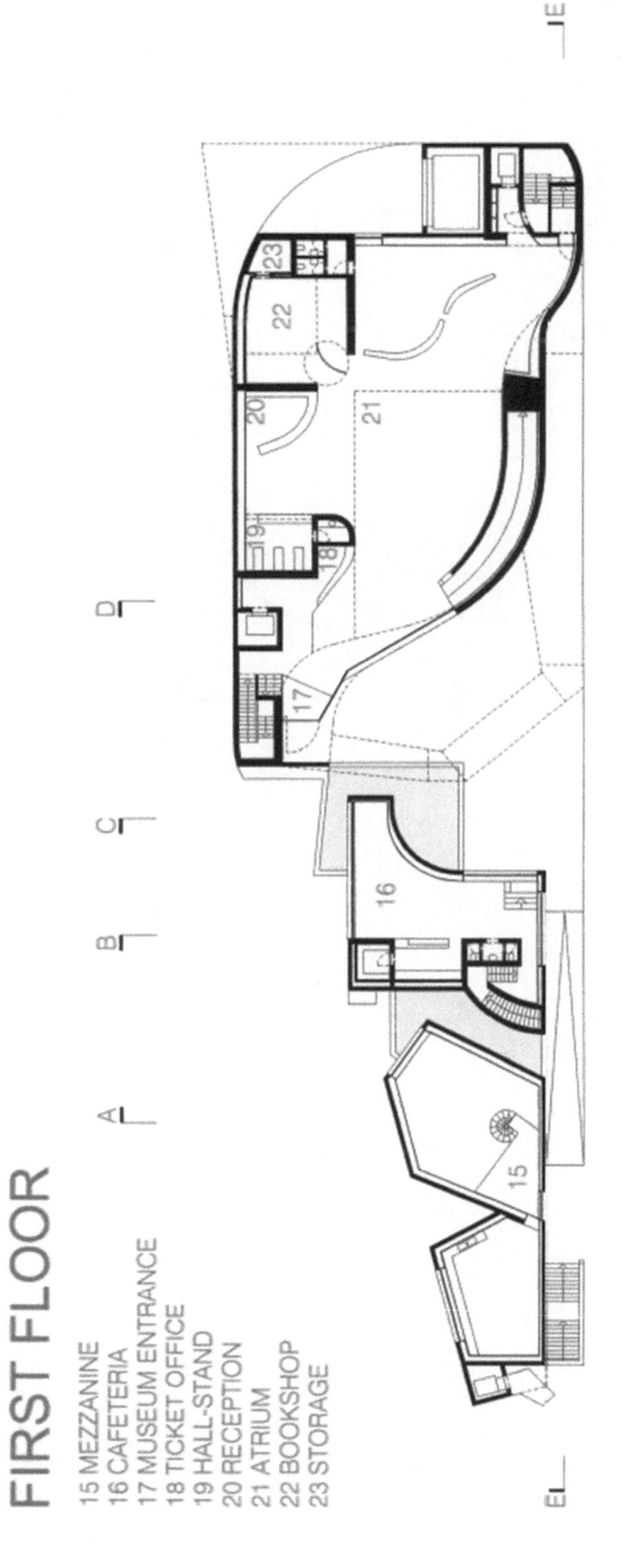


Frist floor plan

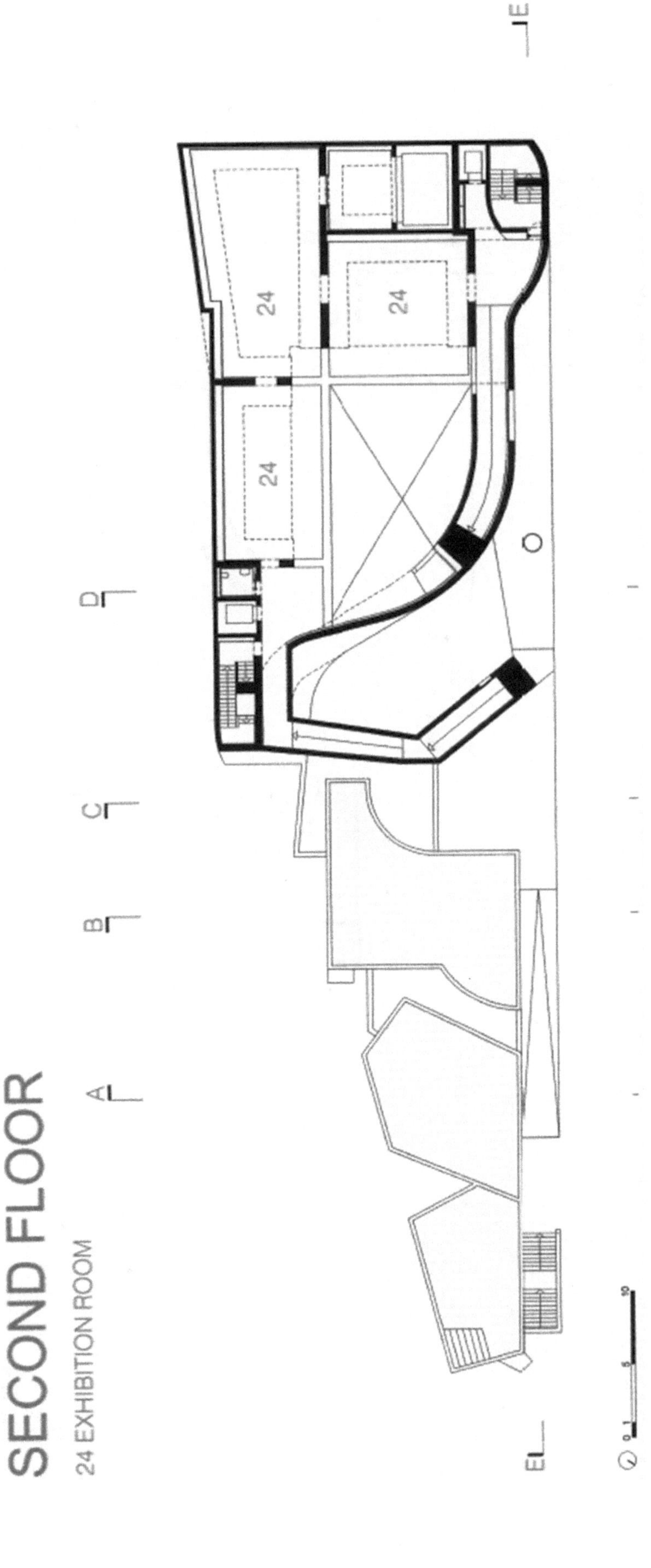


Second floor plan

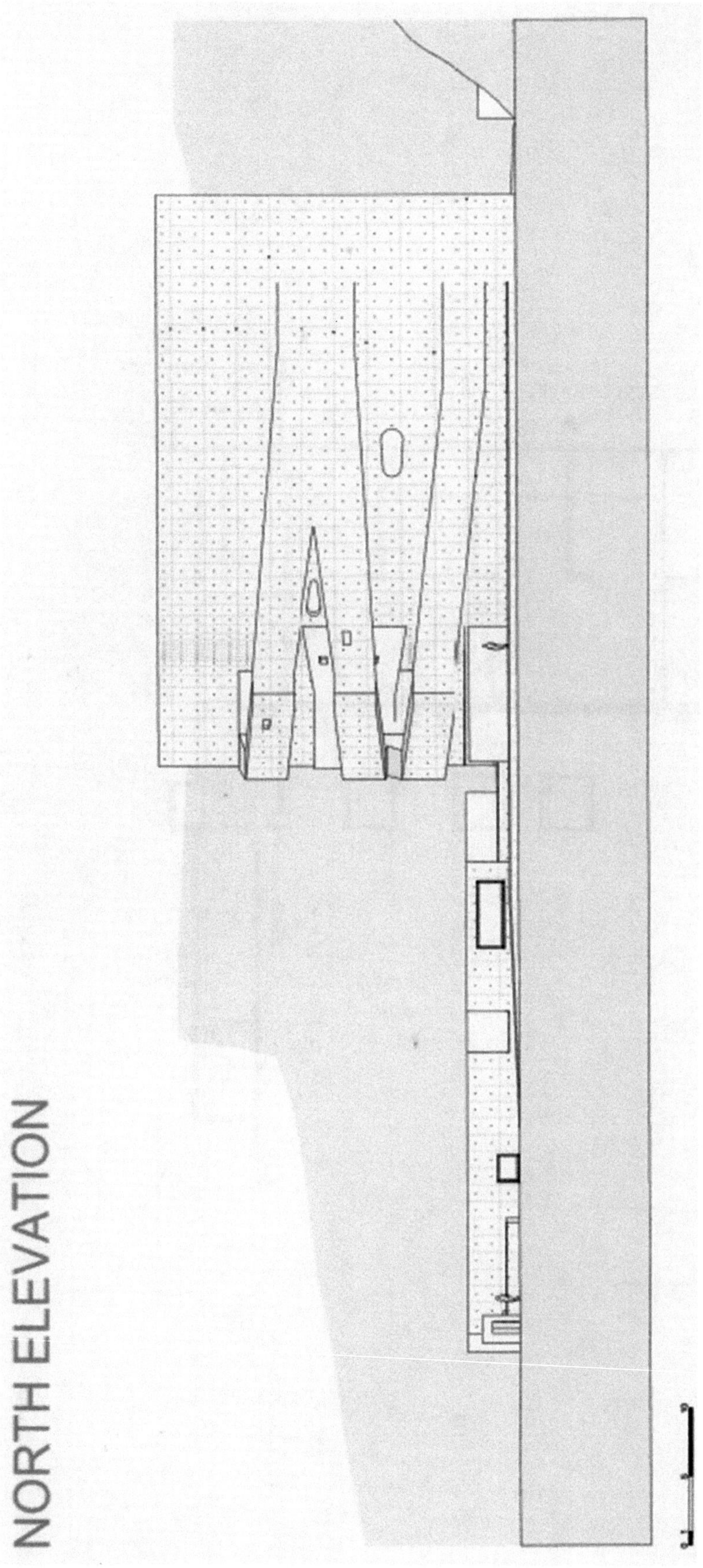


Second north elevation

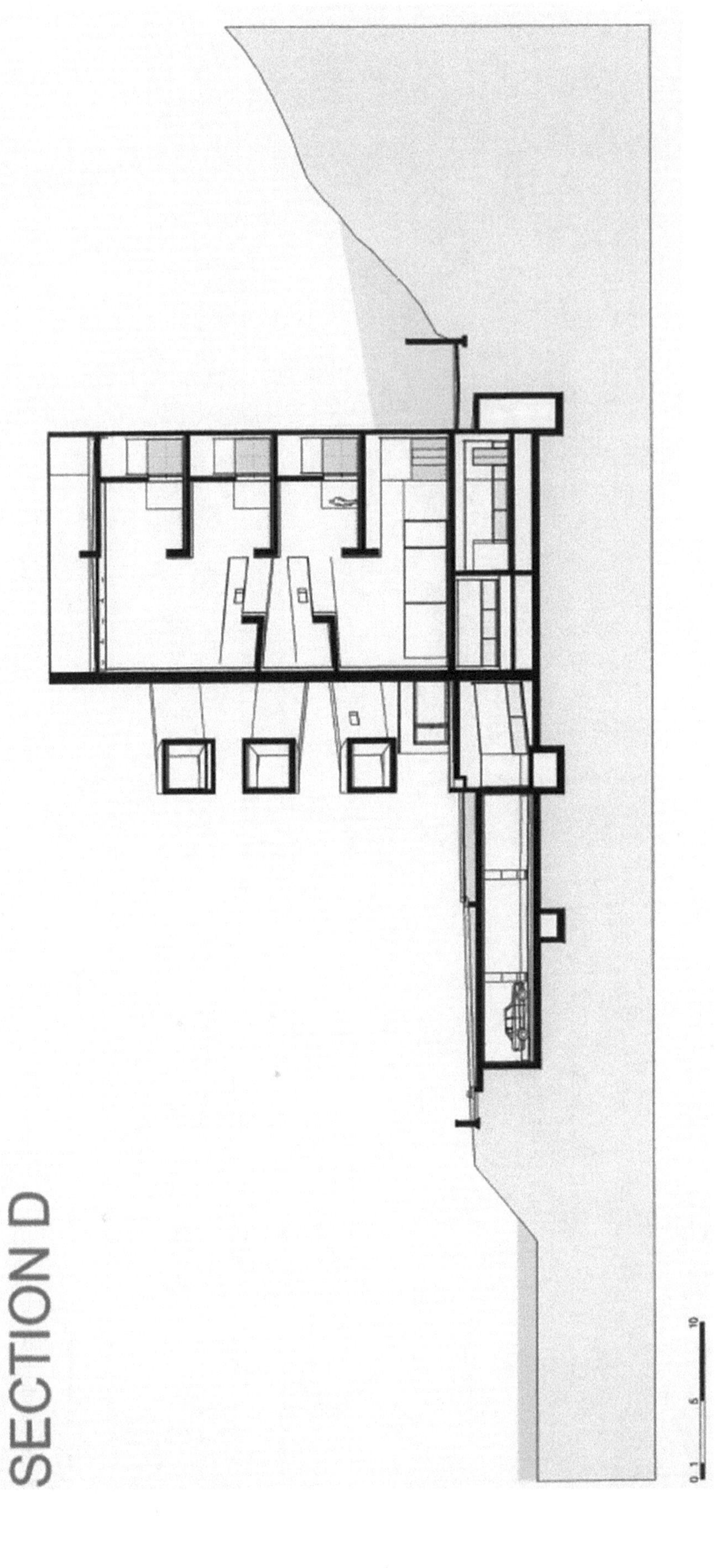


Cross section

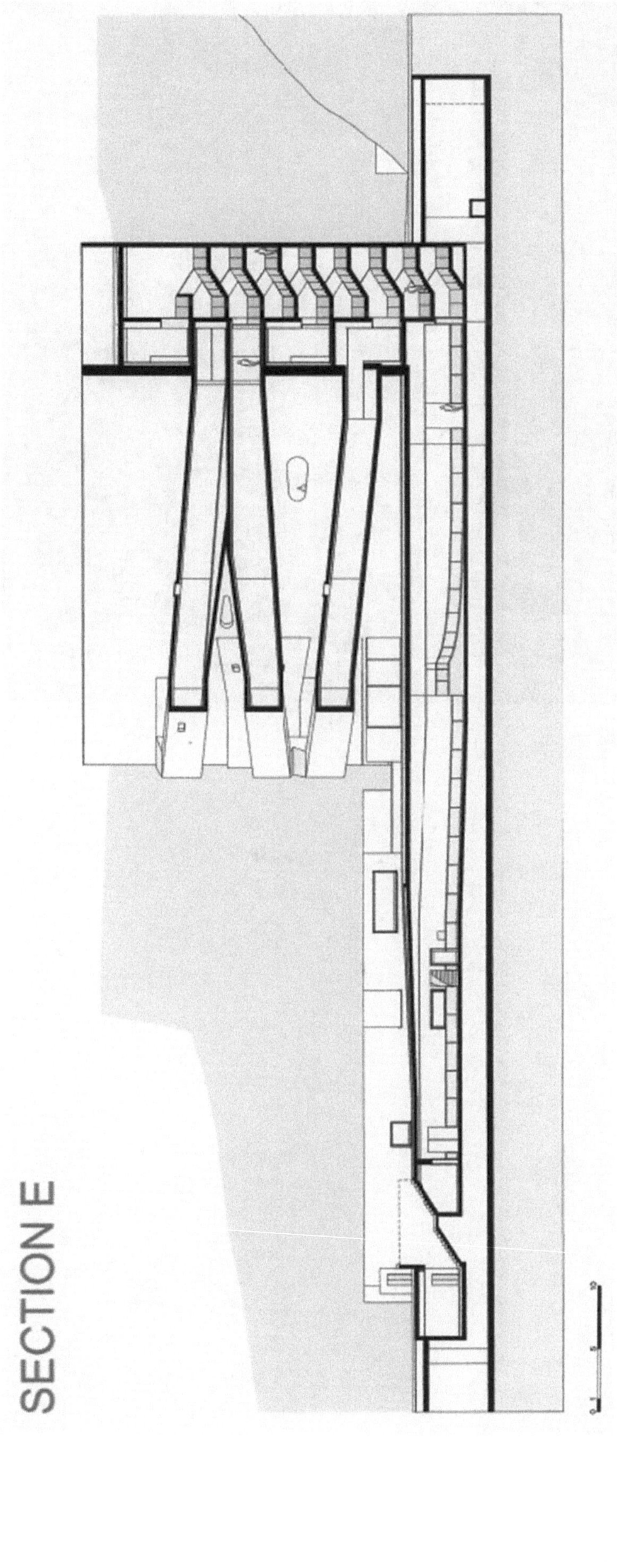


Long section

Street view
Copyright Duccio Malagamba

Internal view of the main entrance
Copyright Duccio Malagamba

# Iberê Carmargo Foundation: Why Was This a Successful Project?

Jose Luiz Canal

I hope that this testimony enables the reader to get to know some of the features that are characteristic of a successful project. I say successful because it was the fruit of 10 years' work in which we have been involved. From the start, I was inspired by the guiding principles underlying the high quality of the whole body of work of Iberê Camargo to coordinate a team that could bring a unique project to fruition on a magnificent and problem-ridden site.

This led to our meeting with the architect, Álvaro Siza Vieira, a true master, known by all of us here, his Brazilian friends, as Architect Siza.

The harmonious and methodical way in which his design team works is extremely smart, always aiming to provide high-level technical solutions for the master's beautiful sketches.

In this sense, Siza's coming on board worked in favour of the artistic quality of the project. It was managed by a small trust led by the artist's widow, D. Maria Camargo; also involved was Dr. Jorge Johannpeter, a high-profile businessman, who has sponsored various arts initiatives, such as the Mercosul Biennial, and who is very involved in quality programmes; Justo Werlang, an astute collector; Eduardo Haesbert, Iberê's assistant; and Martins, the executive director with whom I established a lasting relationship of great empathy.

I mention this small group only to demystify the idea that complex tenders, intellectual and complicated briefs, and rambling discussions are required to achieve an exceptional result. In this case, this small group was able to make quick decisions and naturally recognised an ideal partner in Siza. They gave him their full and constant support, and understood his working methods, pace and the conviction of his ideas.

An atmosphere of clear harmony was therefore established for the development of all the stages of the project, which I will now briefly describe.

This harmony was achieved because we were able to clearly understand who was really the master, the huge difficulties involved in creating and then developing a prototype project, and the tropicalisation of his technical solutions here in Brazil, and also because we kept tight control over the budget.

So, when we deal with a project as ambitious as this one, it is essential to understand the pace, in order to avoid unnecessary stress and maintain a steady rhythm.

The project was divided into four stages:

- 1998/1999 – competition for ideas
- 1999/2002 – adjustments to the brief, including an underground car park, after Siza's and his team's first visit to the site, up to approval and granting of building permission by the City Council and the support of cultural agencies (patronage law)
- 2002/2003 – preliminary project with emphasis on competition for the execution of the infra- and superstructures.
- 2004/2007 – final design, which was developed in tandem with the three work stages detailed below.

This division was very important, because it enabled the whole team to become focused and direct its efforts towards overcoming the huge siting difficulties, the execution of a unique white concrete structure and the appropriate finishings for a wide number of installations, according to Siza's minimalist pattern. In short, we wanted to show everyone that it was possible!

As regards the competition for ideas, I vividly remember the first meeting with Architect Siza in his old studio, particularly my nervousness and his deep respect for the work of Iberê Camargo, obvious from his careful reading of the catalogue of the artist's work, which we had previously sent to

J. L. Canal (✉)
Architecture Department, Federal University of Rio Grande do Sul, Porto Alegre, Brazil



B. Rangel et al. (eds.), *Museum Technology and Architecture*, Building Research: Design, Construction and Technologies,
https://doi.org/10.1007/978-3-319-76172-5_2

him, together with the topographical survey and photographs of the plot, the river, and the sunset that is our pride and joy. I also remember a conversation with Dr. Jorge about his expectations that the building should have a strong character, even to the point of becoming a concrete and steel symbol, a solution that provoked emotion and surprise. In March 1999, I received a phone call informing me that Siza's envelope had arrived; Martins and I opened it, recording this moment in a photograph. The image of the volume electrified me, the drawings with middle plans to make it easier to read the ramp scheme, the white concrete, the façade openings and a delicate, synthetic description. It was a message from Siza, showing something exceptional, with the hallmark of a masterpiece. At this moment, a second penny dropped; we were now even more committed to the challenge of lifting a high quality project off of the paper and overcoming huge practical difficulties in Brazil; but with a dreamer's heart and hard work, we could overreach ourselves and build this gift that architect Siza was giving to Brazil, the land of his ancestors, who lived here, in Belém, at the other end of the country.

After he delivered his preliminary design, we spent a few months concentrating on clarification meetings with supporters of the ICF, and then, in late 1999, we decided to go ahead, which meant flying Siza and his team over to Brazil to see the site. In May 2000, the team landed in Brazil after a long flight, even staying overnight on Governador Island – Rio de Janeiro, in a place that had no windows that would allow them to see Christ the Redeemer. Can you believe it?

But the arrival on site was prepared with the help of Iberê Camargo, the bright sun of our "May Summer," the river like a mirror, the woods and the vista of the location providing the pathways around which we would be building. And, of course, the road was full of cars and buses hampering our ability to change viewing angles.

From this initial visit, it became clear that the siting for the work was perfect; we would need to move the road… and build an underground car park, honoring Siza's wish "not to touch the woods." Incidentally, it must be said that he was right, and today, the woods are dense, thanks to the light barrier that the building creates from the cold south winds.

Two years (2000 to 2002) ensued during which the project was adjusted, with the technical areas adjacent to the car park being repositioned in order to preserve the hillsides even more, since we were talking about a 6 m excavation and everything could crumble in strong rain. Fortunately, the quality of the reasonably preserved granite enabled the subtle placement of a technical area by GOP, which distributes all the utilities, houses the air conditioning plant and the water tanks, and also protects the building in case of flood and/or natural disasters.

We were able to accompany the engineers, Jorge and Bessa, on their building site inspection visits, in the course of several visits to Portugal. Visits to works in Santiago, Famalicão and Barcelona were fundamental for us to get to know various details of Siza's construction vocabulary. We made important visits to Rem Koolhaas's Casa da Música in Oporto and witnessed the difficulties and solutions in building walls laid out in white concrete.

After a few months, we were ready to go, since the siting of the building had been finalised. Once again, the difficulties stimulated Architect Siza to come up with a brilliant solution for the ensemble of preserved hillside/busy thoroughfare/narrow riverbank, while observing strict accessibility standards.

But we would still have to win approval from the city for what was a very unique project, involving very specific issues and a sizeable volume of public land. In today's Brazil, this is rare indeed, because"fears" make the combination of the public and private sectors an uneasy one.

On the other hand, we were voicing the will of the whole community to house the invaluable collection of about 4000 works that D. Maria Camargo had collected throughout many years alongside Iberê. Museums elsewhere in the country would have added considerably to their prestige by having these fantastic works in their collections. But Iberê also wanted his collection to remain in Porto Alegre, in Gaucho land.

Using a local model maker, we created a model illustrating the solution, which clearly showed the underground car park, as well as the shifting of the road that would reclaim more pavement space and correct the distorted bend to increase road safety, which was an important improvement for the city. This model has been with us throughout the project, and was very important in enlightening the technical officers and politicians from the City Council, the neighbours and the thousands of students who visited the site. We worked on development of the project for 2 years, receiving the support and input of a Technical Committee selected by the relevant Secretariats and Departments. This atmosphere of mutual understanding and respect was established from the outset, when Siza delivered a lecture at MARGS, in 2000, for the City Council's technical officers and other official guests.

With the help of these specialists, we found legal solutions, at times quite complex ones, and were granted official building permission in 2002.

At this point, we were visited by Siza, who had been in Salvador for a conference. We took the opportunity to organise the event for cutting the first sod. This was a remarkable occasion. Following a thunderstorm, a very meaningful ceremony took place in which everyone could glimpse the imminent start of the works. In his speech, Siza said that the rain would bring us luck!

A few months later, to add to our happiness, we got a phone call from an emotional Siza to say that he was on his way from Barcelos to Venice, because the project had won the Golden Lion Award at the Venice Biennial.

I confess that, when I put the phone down, I cried, because I never dreamt we could get this far: just to have a project by Siza on Brazilian soil was already a great achievement, but to receive a prize that was normally unheard-of in the Americas, really… This emotion brought with it a sudden sense of heightened responsibility; we couldn't spoil the whole project with a mediocre construction. What a shame that would be!

The award was extremely well-timed and helped enormously in getting the project off the ground, because it became easier to obtain speedy final technical approvals under state and federal patronage laws. These laws are essential for setting up a fund-raising programme and starting the building tender launch procedure.

But, once again, everything worked in favour of the project.

We set up a group of senior technicians to support the detailed Mechanical and Electrical projects. They all acknowledged the worth of the project, and were committed to it, focusing on making solutions viable within a real tropicalisation of the specifications, in an effort to bring on board new state-of-the-art technologies that were not actually available, but were nonetheless feasible.

We knew that we needed to serve as the perfect platform for launching new products and constructive systems while, most importantly, observing the strict specifications defined, even if it meant that construction would take longer.

Another important decision was to divide the works into three stages:

- phase 1 – Infrastructure – up to the 6.40 m elevation height
- phase 2 – Complete superstructure
- phase 3 – Installations and Finishings

We conducted some research and selected five companies with the appropriate technical expertise and a genuine interest in the project's issues to vie for the construction contract.

Camargo Correa, a Brazilian company known for large-scale works of high complexity, was the winning bidder. Their technical proposal demonstrated that, beyond the fact that they had their own white concrete plant, they were also more technically prepared.

Their engineering staff, highly specialised in PERI systems, were a key factor in ensuring the quality of the structure, as were the concrete reinforcement specifications, the installation of a small concrete mixing plant on site, etc. This new technical partnership brought added value to our team.

Another important technical partnership that must be mentioned was with the teachers and researchers from the Federal University of Rio Grande do Sul, through LEME, a materials laboratory, for the development of samples to define the tropicalised white concrete mix, and LPM, a mineral research laboratory that prepared the entire geotechnical excavation project.

One of the most important documents issued by local consultants was an innovative virtual 3D model of the soil to be removed. This model helped to dispel "ghosts" that had haunted us in regard to the quality of the rock and the relative difficulties we might face in its excavation. We managed to remove nearly 40,000 m3 without using explosives. The definition of a weaker splitting plane based on the geotechnical studies enabled quick and extremely economical earth removal work. Given the good quality of this earth, it was fully re-used by the Porto Alegre City Council to pave the city's poorer neighbourhoods. This was a real "goal" of the project's function, and why not one of sustainability, a concept so fashionable today?

We decided to break the excavation work into three sectors (museum, museum-side car park and riverside car park).

This division arose due to traffic constraints during the excavation works and the requirements for the slab concreting of the car park's roof. This was divided by a two-part longitudinal joint, so that at least a two-lane carriageway was maintained to allow for the uninterrupted circulation of heavy vehicles. The traffic study demonstrated the need to complement our strategy by building a detour through Taquari Ave.

This period between July 2003 and May 2004 was extremely stressful, because, even with the detour, there were still traffic jams and a decline in the public perception of the project. This was only natural, since the thousands of people who were delayed there for a few minutes on their way home knew little of what we were building. At this point, we decided to publish aerial photographs of the development of the works, so that everyone could get an idea of the magnitude of the project.

This move was vital to ensure that the community would embrace the project as its new icon, rather than challenge it, as happened, for example, with Portzampac's Casa da Música project in Rio de Janeiro.

Throughout this time, we had other challenges to overcome, such as developing the mix and training the selected team in making white concrete. We decided to use the underground floor walls to train the team with our small concrete mixer. Many features of the project were resolved, mostly those related to forms, blending in the same plane at each concreting stage, timing and correct vibration in layers, actual curing so as to avoid breakage, hot-dip galvanized reinforcements, etc.

If we wanted to create an extremely well-made structure, we had to be as well prepared as possible, with a very strict concreting schedule, from concrete generation capacity, application times, and joint sequence strategy to shrinkage control, shape design, cross-hole control, vibration, 7–10-day wet curing, etc.

For this, we followed the guidance of engineer Jorge Nunes da Silva, the GOP coordinator, a highly competent perfectionist, who was totally in tune with Siza, a wonderful partner in the whole process, and is today a bosom friend.

By October 2004, we had concreted and shaped the whole structure of floor 0, as if it were a "ship's hull," similar to those that ply the canal facing the site. Indeed, this is an incredible visual detail: to see the ships coming towards the window and then turning before the building as if they were giving way to our own looming ship.

A new event was then organized that deserved another important visit by Architect Siza and his team. We decided to improvise a full team meeting in our future auditorium, as if it were an auspicious sign. But this time, we were amazed as, in silence, we saw the virtual walkways around the building with Iberê's paintings on the walls.

I had difficulty sleeping during a significant portion of the construction phase for all the concerns that remained: how to make the abutments, how to control the layout of the foundation slab, the lines, etc. To cap it all, we decided that the core structures of the building would be set in conventional concrete, for the obvious reason of the difference in cost between the two concretes. On the other hand, we needed to control the joint better so that the grey concrete would not compromise the outer white concrete wall.

Once more, it was important at this point to phase the work so as to increase focus. We built a model of this phasing, dividing the work into about 150 concretings.

This model was fundamental in establishing, together with the workers, a quality plan that evaluated operational improvements for the next level up. The result was a clear demonstration of improvement, in which higher levels ended up being much more homogeneous.

We decided to start from the flat back wall, together with the grey slabs of the rooms, to gain more confidence, and then proceed to the curving wall. Soon afterwards, we proceeded to the ramps. This was another challenge, because building the ramps (sleeves) meant working with special reinforcement at the junctions, and narrower walls with a high reinforcement density and different layout. We concreted the foundation slab so that the joint stayed underneath and formed a stretcher bond, and then concreted an inverted U to shape the wall and roof of the outer ramps. All of this with formwork stripping time control to avoid stains. Today, I recognise that the additional effort of creating the joint underneath, because of the importance of water tightness and of avoiding the concreting joint on the wall, was yet another correct decision by Siza in favour of the beauty of his project's details. The simplest solution for the process would have looked horrible.

These ramps were long and hard thought-out; the situation brings to mind an episode from my professional life, when I studied the Van Nelle Factory during my PhD in Barcelona, its quality and constructive innovativeness, the reinterpretation of Lina Bo Bardi in Sesc Pompeia, etc.

We were in a 12 month-long drought, which made it necessary for us to control curing even more tightly. After several attempts at manual curing with blankets, we solved the problem in a simple way with the help of a garden watering system, which, besides being easily installed, had the added advantage of a softer jet that did not stain the concrete.

At this point, it was important to follow the instructions given by engineer Jorge to the local structural designer, Fausto Favale, with a proper reinforcement delivery schedule on site that left no room for storage. For this, the contribution of Gerdau Armafer was very important, processing the cutting/bending of the nearly 1500 tonnes of steel and sending them for galvanising and then on to the site, in a perfect system that worked extremely well. I say extremely well because there was little rehandling and/or few losses of reinforcing steel bars; we were even forced to adjust the machinery to deal with the very large steel bar radii demanded by the design.

After his previous view of the site during the concrete reinforcement stage in October 2004, Siza returned via Recife, in December 2005, this time after an interesting visit to the Brennand Museum, where we were looking for special ceramic tiles for parking.

I will always remember the tour I made with Siza and Pedro at sunrise, after having endured a ghastly flight with many stops and much backache. But the emotion of seeing the master before his work is an unforgettable image!

At this point, our "canteiro" (building site, or "estaleiro," as they say in Portugal… indeed, throughout these years, I have learned many technical expression in European Portuguese that leave my colleagues mystified whenever I use them. Sometimes it's hard to switch back…) But to continue… our "canteiro" was completed, about 150 people were working and we had a well-organised office to manage purchases and payments, connected via the Internet with Siza's team. We prepared reports so that the overseas team could follow what we were doing, our work rosters, and what was still undone and would require operational instructions soon.

After the technical specifications, it was fundamental to go over prototypes of the parts of the building to assess problems with the abutments; in Siza's case, on the same level, which increased difficulties, although the finished result is infinitely superior to the usual solutions.

A new event followed, with a visit to the site by D. Maria, D. Jorge and other friends. This visit was very important, because it led to Siza and his team receiving the acclaim they deserved for the beauty and sensuousness of the structure. In some ways, it evokes the shape of a saxophone, of which Architect Siza is a great fan in his leisure time. Many ties were established, but we still had a great challenge to face: that of fitting and facing everything.

In our case, things needed to be done in a different way, because the façade's structure was complete, and thus we had to build from outside in, in the following order: rainwater drainage, thermal/acoustic insulation, ancillary installations and drywall facing according to usage criteria.

We decided to take on direct management of the whole works, with the support of Camargo Corrêa's fantastic team, led by Master Rocha for the structure (stages 1 and 2) and Master José Leandro for the installations and finishings (stage 3). We also had effective support for document control and the preparation of detailed dimensioned plans from Architect Camila, who supported the interpretation, or rather the recipe for the execution of the work. All the while, we had the support of a full-time team of surveyors for pegging out and controlling everything, special consultants for more complex disciplines, a purchase and payment structure and a strong emphasis on the safety team. Safety was indeed a fundamental aspect, and we scored a record when we completed 1000 days without a single accident at work, a much celebrated moment with a big barbecue in the foyer for the labourers involved.

At this point, we divided the third stage into four phases:

3A – Curing
3B – Envelope closure / Fit-out
3C – Finishings
3D – Equipment, furniture and the opening.

Once again, this division was essential so that we might focus, in an organised way, on the various utility networks. Layer upon layer, we were completing the building, always following prior work with qualifying prototypes from suppliers so as to select the subcontractor teams. These prototypes were evaluated primarily for the quality of the results achieved within predefined values.

We negotiated with approximately 30 different companies for the final services, e.g., waterproofing, insulation, hydraulic, electric and special installations, fillings, and the largest package of this stage: the air conditioning systems. The result of engineer Bessa's detailed design, it encompassed various independent solutions, managed by a robust open protocol automation system using Lon language from Johnson Controls. The system allows for future upgrades, so the building will not become obsolescent.

After long negotiations, the company Heating Cooling from São Paulo was contracted, since their background included large-scale works, including complex installations at Petrobras oil rigs.

We managed to procure the package specified in the design, with the support of the manufacturers and also foreign suppliers, who were partners with Siza's office, as in the case of Uponor radiant mats.

The late arrival on site (stage 3B) of the air conditioning systems was important for us to be able to maintain control over all work streams, supervising each detail of the laying of the pipe and hydraulic network. We could not afford to make mistakes, for fear of compromising the thicknesses specified in the architectural design.

We made small adjustments and fine-tuning in the technical area. This technical area was gradually filled with system upon system, and was quite well-organised. This is an essential factor in facilitating maintenance.

Now, we were at the same time dealing with the purchase of finishing materials. The marble was due to arrive in port in 15 containers. It was of various thicknesses; as the containers were opened at the contracted marble yard, we were thrilled with the beauty of the material.

We had more trouble with the purchase of the flooring, as we were forced to change both supplier and type of material. We started with guatambu, which is rather light, but found that it was not produced in long boards, because it comes from a small Paraguayan tree. We resorted to a different, larger supplier, who recommended flamewood to us, a light-coloured material supplied in long boards. With this material, we were able to satisfy the continuous layout that Siza wanted, which did not provide for connection pavements between two sides.

We must mention the difficulties of facing walls, especially acoustic ceilings with perforated gypsum plates, which were covered with fabric and water-based permeable paint, not forgetting the assembly of the slender factory-made skylights.

We carried out a new topographic survey to adjust minor differences and avoid defective couplings and/or projections, sheer imperfections that we could not tolerate.

Siza, through his most direct associate Pedro Polónia, was always present, and everything was specified and defined in detail so that abutments were under control.

In the end, the final result of the finishings left me astounded at the sensuousness of the form, the light that streams through the windows, the abutments that make up the composition of the minimalistic facing.

We cannot describe the final stage of the installations and finishings without mentioning the aspects that actually relate to sustainability. I have always been a little wary of theoretical discussions about these issues of compromising with the environment. Particularly as I come from a beautiful country, albeit an irresponsible one as far as the protection of its abundant natural resources is concerned.

In our case, the return of the major investment was in the reflection of the white concrete, the openings, full insulation, the re-use of rain water, the complete wastewater treatment system, the reintegration of the José Martí Plaza as a park surrounding the building, and many other qualities, beyond the initial design features, which make this work an exemplary case study in sustainability.

Today, our constant concern is, above all, preserving what we spent 10 years building so painstakingly. For this, a 7-strong Preservation Committee was set up. This Standing Committee will be permanently responsible for supervising the building's operation and maintenance. It reassures me to know that future generations will take care of their icon.

A deep sense of emotion will always be with me when I look at the building and see thousands of schoolchildren experiencing the magical spaces created by Siza. And we are also proud of having made it possible: an internationally renowned building in terms of construction quality and sustainability. A house that lives up to the ambitions and work of Master Iberê Camargo.

And, lastly, I must remember the most important thing: the feeling of elation of the whole team, touched by the acclaim from the general public at the grand opening ceremony. Meanwhile, we all became orphans of the project, of the daily phone calls, the emails, the conversations with Siza on Sundays, of Serralves, of Boavista in Porto, of the "caipirinhas," the barbecues at Barranco, watching the sunset over the works during a cigarette break, the last of which certainly stands as a fundamental representation of Siza's pace.

# The Efficiency of Air Conditioning Energy Systems in the Iberê Camargo Contemporary Art Museum – Rio Grande Do Sul – Brazil

Raul Vasconcelos Bessa

The mechanical engineering project designed by the company GET – Gestão de Energia Térmica Lda. (url: www.get.pt) – not only encompassed the design of the most effective systems, in terms of energy, for responding to the thermal needs of the building arising from its envelope and its internal gains, but was also developed from an integrated perspective of energy streamlining while always maintaining good interior air quality.

The main objective defined for the HVAC project for the Iberê Camargo Museum was the implementation of an exemplary project demonstrating the use of efficient and environmentally friendly energy, with a view towards a decrease in consumption.

For this purpose, it was paramount at the design level to restrain the dimension (power) of the air conditioning systems so as to avoid excessive oversize by making a proper selection of the primary energy providing systems and Rational Utilization of Energy options, assessing their economic viability, and thus minimising the respective energy consumption.

Using these guidelines, therefore, a set of energy consumption rationalisation measures were taken, such as heat recovery from the chiller condenser to achieve terminal reheating of the batteries of the air treatment units for interior relative humidity control, the use of free cooling of the air treatment units, the use of variable flow and high energy efficiency pumps for distributing both heated and refrigerated water to terminal equipment, the use of radiant air conditioning in the outer circulation tunnels between floors, and the implementation of a centralised technical energy management system.

The treatment of ambient air is achieved by the air treatment units that ensure interior thermal comfort, although there is independent control of the blowing temperature for each space and there is provision for terminal reheating batteries fed by the chiller condenser water.

In addition to these measures, power splitting in refrigerated and hot water generating equipment was taken into account, as well as the use of more efficient electrical equipment (class A). In particular, the adoption of energy and environmental efficiency criteria was promoted in the purchase of the main energy-consuming equipment.

Free cooling is a technique based on the use of outside air to achieve cooling, in this way decreasing or eliminating cooling requirements.

An air free cooling system, like that installed in the air treatment units, recovers fresh air from the outside when its temperature and relative humidity (i.e., outside air enthalpy) are lower than the values selected for inside conditions.

Through the use of free cooling, energy consumption in the building is reduced, because the chiller is disconnected while the free cooling is working.

Thermal energy storage was used for ice storage cooling and good preventative maintenance practice was implemented.

Storage with ice is based on the high value of the latent heat of melting water, around 355 kJ/kg at 0 °C. A reduction in the accumulation volume is achieved of up to 25% of what would be necessary for accumulation with refrigerated water for the same amount of accumulated energy.

These systems require low evaporation temperatures in the refrigerating machine, on the order of −12° to −5 °C. It is normally necessary to use a glycol-based cooling fluid to prevent water freezing in the cooling circuits.

A solution with 25% "ethylene glycol" is a current solution for the secondary cooling fluid. Depending on the system's dimensions and usage, the use of an exchanger may be preferable so as to prevent glycol water from circulating throughout the whole distribution system.

The accumulation of thermal energy for ambient cooling consists, therefore, of the temporary storage of thermal

R. V. Bessa (✉)
GET Gestão de Energia Térmica, Lda, Porto, Portugal
e-mail: raulbessa@get.pt



B. Rangel et al. (eds.), *Museum Technology and Architecture*, Building Research: Design, Construction and Technologies,
https://doi.org/10.1007/978-3-319-76172-5_3

energy at a low temperature during off-peak periods for later use in periods of higher energy cost.

Thermal storage technologies in general, under which thermal accumulation in support of cooling processes within HVAC installations falls, enable system operation during off-peak tariff times (night time), when electricity is cheaper, for later use during periods of higher energy cost, reducing the value of the power taken, essentially at peak times.

Thermal energy for ambient cooling is thus stored during this period and is later channelled into the building during working hours.

The accumulation of thermal energy for ambient cooling, in view of its technological features and operational methods, presents various advantages, as described below:

1. An increase in chiller efficiency due to preferential operation at rated speeds, which also contributes to the life span of the equipment.

On the other hand, the preferential operation of cooling machines during the night will enable their FOC to be increased, since they operate at lower condensation temperatures. In this way, and depending on the optimisation degree of the installation, it is possible to decrease energy consumption.

2. A decrease in the power of refrigerating equipment and lower simultaneous operation with other equipment, contributing to a decrease in the power taken by the installation.
3. The possibility of use of high $\Delta T$, which will allow for lesser flows, with a resulting reduction in load losses and smaller duct dimensions.
4. A reduction in the electricity bill, in terms of both power taken and energy consumed, since the power taken decreases and, with it, the power contracted can also be reduced. The energy component decreases as well, due to energy consumption shifting from peak times to off-peak times.
5. Reduction of noise levels produced due to higher operational stability;
6. In the case of the use of iced water tanks, these may also be used as a water reservoir for fire fighting;

This solution, therefore, enables electric energy consumption to be reduced because the energy costs of system operation are lower, also leading to a reduction in the cost of power, thanks to the operational time-shifting of refrigerated water generating plants.

The disadvantages of these systems are essentially concentrated in higher start-up costs and the need for available space in which to install the ice storage tanks.

However, this increase in cost is normally recouped within reasonable periods of time due to the resulting savings.

In this way, this technology applied in support of the large air conditioning system installed at the Iberê Camargo Museum represents an interesting solution that was envisaged by the designers and the directors of the building.

The adoption of the use of more efficient artificial lighting equipment and bulbs was promoted, through the accrual of a credit, by demonstrating the utilisation of the most efficient categories of these energy consumers, according to the respective energy labelling certificates.

An estimate of the building's energy consumption was made using detailed computer methods, which enabled its global consumption to be predicted, under the anticipated rated operating conditions of the building, with the main aim of ensuring that the building's consumption does not exceed figures that are considered excessive.

However, energy consumption at the Iberê Camargo Museum depends on many factors that go beyond the detailed energy simulations carried out during construction, amongst which are the level of visitor flow, the behaviour of energy-consuming installations, the periodicity of maintenance work, etc.

One of the places where innovative measures were required was the radiant air conditioning system designed for cooling and heating (if needed) of the circulation tunnels between floors that are visible on the building's façade.

In view of the high thermal load of the envelope of the tunnels/corridors visible on the building's façade, for circulation between floors, and the reduced space for installing an all-air-type ambient cooling system, a radiant air conditioning system was favoured, using polypropylene grids rebated into the walls and ceiling of the top floor.

Radiant heating and cooling systems are more energy-efficient than traditional convection systems.

In environments where air speed is low, as is the case with interior spaces, radiant heating and cooling has the advantage that most of the heat transfer from the surfaces of a human body occur in the form of a radiant interaction with the surrounding surfaces.

In the case of these circulation spaces, heating and cooling thermal loads mostly originate through their exterior envelope and structure, whose surfaces, directly exposed to exterior conditions, reach extreme temperatures.

Radiant heating and cooling, i.e., a radiant air conditioning system, is able to maintain a thermal equilibrium inside these corridors of the building through a radiative interaction with the inside surface of the outside walls, allowing them to maintain a constant temperature closer to that for desired interior comfort.

Heat flowing directly onto the radiant surfaces can be immediately absorbed by them, merely increasing the circulation of water flow in them without altering their surface temperature and without requiring the flow or the temperature of the air renovation of the space to be changed.

In view of the high values of both the mass heat and the volume mass of the water, heat is transferred to or from the interior of the building, consuming only about 1/5 to 1/10 of the electric energy that would have been required to achieve the same heat transport by air, due to the glitch energy required to feed ventilators and hydraulic pumps.

Radiant cooling follows the same principles as radiant heating. The heat transfer that is achieved between the space to be cooled and the radiant panels is due to a difference in temperature.

However, unlike in radiant heating, the refrigerated ceiling absorbs thermal energy, radiated by people and objects or a neighbouring apparatus.

The greatest difference between refrigerated ceilings and cooling by traditional convection systems (all-air systems) lies in the heat transport mechanism.

Traditional all-air systems only use convection, whereas refrigerated ceilings use a combination of radiation and convection.

In the case of refrigerated ceilings, the amount of heat transfer by radiation is about 55%, with only the remainder being transferred by convection.

With the refrigerated ceiling system, heat transfer by radiation is achieved through emission of electromagnetic waves from warmer people or objects and apparatuses to the refrigerated ceiling.

Since convection starts by cooling the air in the space, due to contact with the refrigerated ceiling and walls, convection currents are created in their interior that achieve heat transfer from generating sources to the refrigerated ceiling and walls that absorb it.

The Iberê Camargo Museum is a fine example of a low energy consumption building.

# Lighting and Museums

Raul Serafim

Lighting in museums can and should cover a wide range of architectural designs and technical lighting, to be taken into consideration by both those who design them and those who manage them, in order to make the lighting of these types of building more sustainable from economic and environmental points of view, and with a colour temperature and intensity suitable for the works that are illuminated.

There should be controlled, diffused natural light in most buildings, museums in particular, applying some of the techniques that were well-studied in the 1960s and '70s.

Up to the beginning of the 1970s, lighting was a topic that was much debated by various scholars, who applied "solar charts" in architectural studies and developed their projects to make the most of natural diffused light, using well-studied and published techniques, such as those in the 1966 publication, "Daylighting" by R. G. Hopkinson, P. Petherbridge and J. Longmore, considered by many to be a Treatise on Natural Lighting.

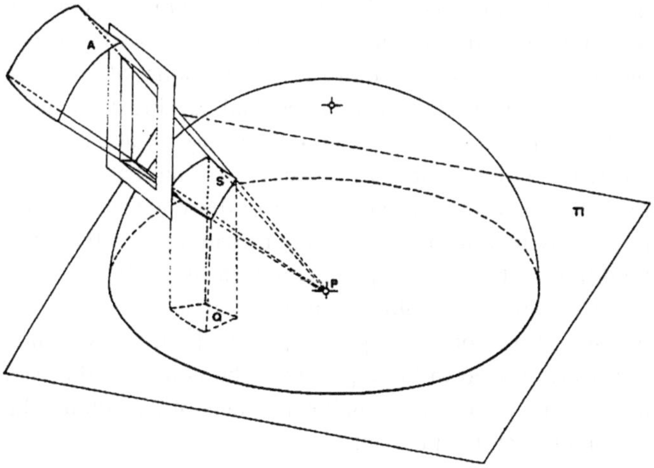


**Fig. 1** Lightning geometric scheme

As a practical example of construction, we can mention the light-flaps located a quarter of the way below the tops of windows that prevent the entry of direct sunlight (damaging to paintings) and project natural light up to the ceiling inside the room with the help of reflectors located above the flap, as in the technique used by Siza Vieira in the Faculty of Architecture of Oporto, where small flaps keep out direct sunlight for most of the day without the help of shutters, thus making use of diffused natural lighting, or as employed in the Serralves Museum through the exemplary use of zenith lighting, as described below.

In both the Serralves Museum and the Iberê Camargo Museum, the architect chose to create entry points for controlled zenith lighting, and also to create light entry points through large vertical wells in the case of the Serralves Museum, the result of a careful study to ensure that there is no penetration of direct natural light.

With the invaluable help of new technologies for controlling diffused natural lighting, such as, amongst others, the Helvar technology used in the Serralves Museum, it is possible to take advantage of the large skylights created by Architect Siza Vieira for the entry of diffused natural light, and so reduce the energy costs for lighting and, indirectly, the HVAC too, without endangering the exhibits or the design of the exhibitions.

In the case of the Serralves Museum, the distance between the interior and exterior skylights was designed and coordinated with the Architect, creating a technical floor (also used for many other functions) where a shutter and a twilight sensor were fitted. The light source used was a fluorescent lamp, equipped with an electronic ballast dimmer.

Through the creation of a specific algorithm, the system receives information from the cell on the intensity of the illumination from the external skylight, compares it with the

R. Serafim (✉)
AFA consult, Vila Nova de Gaia, Portugal
e-mail: Raul.Serafim@afaconsult.com



B. Rangel et al. (eds.), *Museum Technology and Architecture*, Building Research: Design, Construction and Technologies,
https://doi.org/10.1007/978-3-319-76172-5_4

levels of illumination required for a particular exhibition in progress, and then regulates the opening of the shutters so that the light level is as close as possible to what is required.

As night falls, regardless of the time of year, and when the natural diffused light is insufficient with the shutters already fully open, fluorescent lighting lights up at 10% of its maximum level and, slowly, according to the needs of the exhibition, increases the amount of light so as to always maintain the levels of illumination required by the exhibition. This procedure, besides being an effective means of saving energy, increases the average life of each light bulb, thus contributing to an increase in the final sustainability of the solution.

In addition to the rational and optimised use of diffused natural light, it is also necessary to take into account three types of illumination when designing a museum: Ambient Lighting, Accent Lighting and Spot Lighting.

Ambient lighting is responsible for the minimum illumination of the exhibition rooms and corridors, adapted both to the level required for a particular exhibition and for the safety of those moving around and the proper operation of the security cameras.

Accent lighting is responsible for illuminating a painting or sculpture, and should respect the "Law of Reciprocity," described below, so as to be able to capture the spirit of the artist in the eyes of the beholder.

Spotlighting, as the name suggests, is responsible for highlighting detail, for example, to convey emotion, such as illumination to accentuate the image of a woman crying, providing a specific projector to emphasise the tears in relation to the whole.

Another criterion to take into account when lighting a particular painting exhibition is how the author of a work painted it, whether it was painted inside or outside, with a little or a lot of natural light.

This criterion is a matter of controversy among many directors of leading museums and art critics. What is certain is that, through the use of light sources that have a less suitable chromatic rendition, you can change the colours of an exhibited object in such a way that it distorts the work exhibited.

According to André Béguin (author of the world famous "Dictionnaire technique et critique du dessin"), the painter Titian painted the colour of the skin of his characters with a tone of "pig skin" (author's reference) and it glows golden because Titian painted mostly at night, with very low levels of light in his studio. In this case, if the lighting designer were to use the usual levels to illuminate the paintings of Titian, the public would see the skin of his characters in a different way than that which the painter conceived.

In a contract between the painter Grão Vasco and the Bishop of Lamego for the creation of one of the former's paintings, it was written that the diocese would undertake to find a house with sufficient height for the commissioned work, as well as abundant natural light, as instructed by the painter (Information: Dr. Dalila Rodrigues). Returning to the "Law of Reciprocity" and protection against direct natural light, these paintings should be viewed with a reasonable amount of light, because they were done on wood.

Given the two examples above, one of the main issues becomes: How should we view a painting?

… in the same light (colour temperature and illuminance) that surrounded the painter when the work was painted? What about the inherent difficulties, both of obtaining the historical data on the conditions under which the artist created a particular work and the creation of identical conditions inside a museum?

… with the light (colour temperature and illuminance) that displays the exhibited picture at its best? Including the skin colour of the characters in Titian's paintings? Is it right that we should view them in lurid colours? This has happened with the works of many other artists.

... with the light (colour temperature and illuminance) that the Museum Director "thinks" it should have? What training or knowledge of lighting technique does the Director have? And what criterion has been used?

... or, perhaps, under the conditions requested (required) by the Insurance Consultant? In this case, it is usually based solely on the "Law of Reciprocity" and the "x lux hours/year" ratio as used throughout a year.

These issues are very relevant and difficult to answer, because, during design, the pieces that will be exhibited in a particular room are often not known, or perhaps there is still no museum director or even experts on a particular painter at hand, and by the time such folk do arrive at the museum, the work on it is either well under way or has already been completed. All these issues require a greater amount of effort and research on the part of the architect and lighting designer.

To light a work of art is simultaneously to take care that the work is not damaged by excessive illuminance or UV rays, and to make sure that the eyes of those who see it do so in the way that the artist intended, providing it with the best rendition of the colours applied by the artist to the work.

As we know, the colour temperature that our eyes see is a mixture of the colour temperature of the light that we direct at a work, mixed in large part with the colour of the paint used in that work, and a small percentage of the colours that surround us at that moment.

How can we achieve the best and most appropriate colour temperature, or, in other words, what light source must we use so that the human eye sees the colours as faithfully as possible? What techniques should we use?

One idea is to use the Colourimeter Method, linking a colourimeter to a PC and illuminating the painting with a known colour temperature T (K). The colourimeter determines the correlated average temperature of the source represented in the painting Tp(K).

If Tp has the same magnitude as T, this is the best Colour Temperature for the painting in question. If it does not have the same magnitude, we must go on substituting "T" until we reach an approximate magnitude.

If it is not possible to conduct this sort of trial, a quicker, normally less scientific way is accepted for works painted up to the mid-nineteenth century:

- 4000 K for the sun (predominantly radiation, of yellow reaction);
- 5800 K for the Zenith (predominantly blue radiation, with low saturation and high luminance);
- 2700 K for Dusk (reddish orange radiation, with high saturation and low luminance).

In even more general terms:

- Temperatures between 3000 K and 4000 K are used for paintings that portray interior scenes;
- Temperatures between 4500 K and 6000 K are used for paintings that portray open air scenes.

The application of the type of light source is essential for achieving good colour reproduction (CRI). For example, an incandescent lamp has a CRI of 100, whereas a Low Pressure Sodium lamp has a CRI of 0. Last generation fluorescent lamps have a CRI greater than 90, and a minimum CRI greater than 80 is considered for paintings.

When pieces of high artistic and/or economic value are illuminated, other concerns must be taken into account, which we will mention briefly:

- Wave/particle duality

Light is composed of photons, which have an associated energy: $E = h \times f$.

in which:

$h$ = Planck's Constant = 6,6256 × 10 – 34 (J × s);
$f$ = radiation frequency (Hz)

The frequency and the wavelength (λ) are inversely proportional to the speed of light $f = c/\lambda$ $E/h = c/\lambda$ $E = c \times h/\lambda$.

Therefore, the shorter the wavelength, the greater the energy (illuminance), which implies GREATER DETERIORATION.

What do all of these formulae mean? Ultraviolet (outside the visible spectrum) and blue violet shades cause more damage to exhibited works.

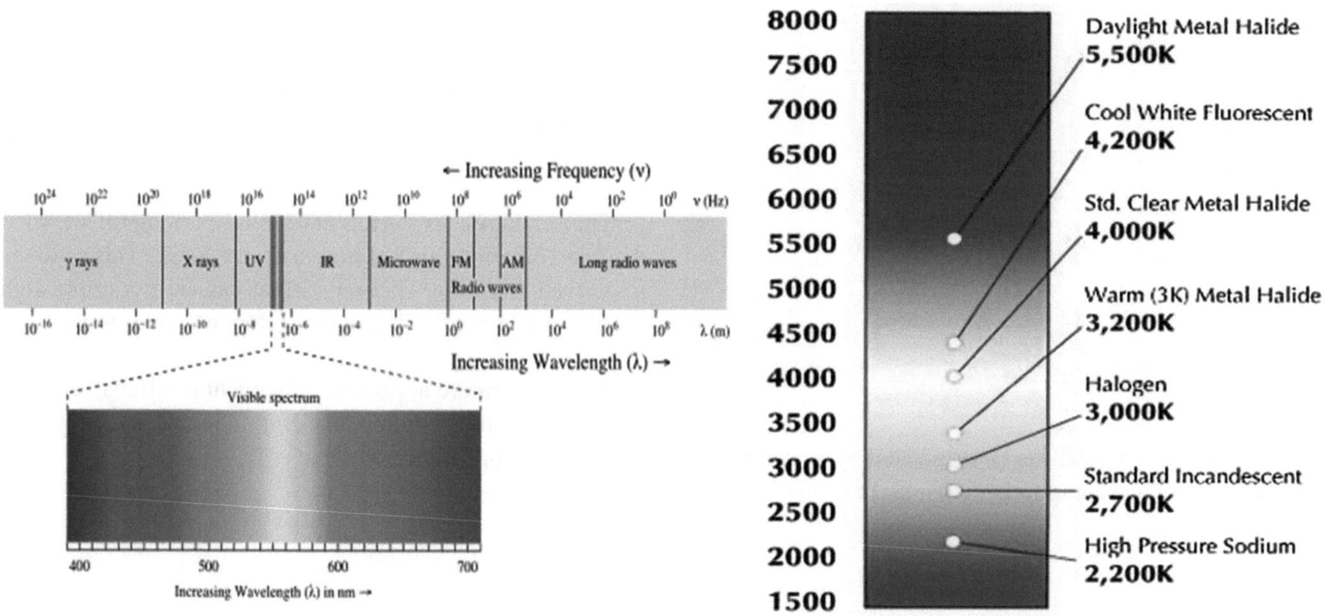


**Fig. 2** Light spectrum

The study and application of the wave/particle duality effect leads us to the following table of Relative Degradation $D_\lambda$ originated by a unit of radiance for various wavelengths:

$$Vpchem = \frac{\int_0^\infty Fr\lambda\, D\lambda\, d\lambda}{\int_0^\infty Fr\lambda\, V\lambda\, d\lambda},$$

in which:

$V_p$ = Damage or Deterioration Factor and
Fr $\lambda$ = Spectral radiant flux (W/nm)

| λ (nm) | $D_\lambda$ | λ (nm) | $D_\lambda$ |
|---|---|---|---|
| 300 | 7,75 | 540 | 0,012 |
| 320 | 4,50 | 560 | 0,007 |
| 340 | 2,63 | 580 | 0,004 |
| 360 | 1,45 | 600 | 0,002 |
| 380 | 1,07 | 620 | 0,001 |
| 400 | 0,66 | 640 | 0,0005 |
| 420 | 0,37 | 660 | 0 |
| 440 | 0,20 | 680 | 0 |
| 460 | 0,12 | 700 | 0 |
| 480 | 0,065 | 720 | 0 |
| 500 | 0,037 | 740 | 0 |
| 520 | 0,021 | 760 | 0 |

The Average Sensitivity Factor of the human eye for various wavelengths is shown in the following table:

| λ (nm) | Vλ | λ (nm) | Vλ |
|---|---|---|---|
| 300 | 0 | 540 | 0,954 |
| 320 | 0 | 560 | 0,995 |
| 340 | 0 | 580 | 0,870 |
| 360 | 0 | 600 | 0,631 |
| 380 | 0 | 620 | 0,381 |
| 400 | 0,0004 | 640 | 0,175 |
| 420 | 0,004 | 660 | 0,061 |
| 440 | 0,023 | 680 | 0,017 |
| 460 | 0,060 | 700 | 0,0041 |
| 480 | 0,139 | 720 | 0,0010 |
| 500 | 0,323 | 740 | 0,0003 |
| 520 | 0,710 | 760 | 0,0001 |

The Relative Degradation $D_\lambda$ originated by a unit of radiance for various wavelengths is shown in the following table:

| λ (nm) | $D_\lambda$ | λ (nm) | $D_\lambda$ |
|---|---|---|---|
| 300 | 7,75 | 540 | 0,012 |
| 320 | 4,50 | 560 | 0,007 |
| 340 | 2,63 | 580 | 0,004 |
| 360 | 1,45 | 600 | 0,002 |
| 380 | 1,07 | 620 | 0,001 |
| 400 | 0,66 | 640 | 0,0005 |
| 420 | 0,37 | 660 | 0 |
| 440 | 0,20 | 680 | 0 |
| 460 | 0,12 | 700 | 0 |
| 480 | 0,065 | 720 | 0 |
| 500 | 0,037 | 740 | 0 |
| 520 | 0,021 | 760 | 0 |

- Illuminance and Exposure Time = Law of Reciprocity

The damage caused by Light depends on the exposure time (lux × hours/year).

In the study of the Law of Reciprocity, art is divided into three categories within the scientific community:

Category 1: Objects very sensitive to Light;
Category 2: Objects sensitive to Light;
Category 3: Objects insensitive to Light.

Within these categories, the average illuminance recommended for each is as follows:

Category 1: 50 Lux;
Category 2: 100 to 150 Lux;
Category 3: The recommended maximum for display aspects (300 Lux is accepted as reasonable).

Another factor brought to prominence by the Law of Reciprocity is the Discolouration Ratio, given by the expression

$$F = Cte \times Vp \times E \times t$$

This Ratio relates the amount of light falling on a work to the loss (deterioration) of colour in the work. This outcome is derived from the Thermal Effect caused by strong light sources on the works due to the following conditions:

- The temperature accelerates chemical reactions, due to an increase in the agitation of atoms and molecules, resulting in the disappearance of some colours;
- The surface humidity reduces.

Applying the "Law of Reciprocity," we present the following possible examples:

- Canvasses, Tapestries, charcoal and pencil: exposed 8 h/day to an average 50 Lux;
- Watercolours and water-based inks: exposed 8 h/day to an average 150 Lux;
- Oils and synthetic paints: exposed 8 h/day between 200 and 250 Lux.

Another factor to take into account in regard to illumination in general, and particularly in museums, is direct and indirect glare. Direct glare is normally caused by the direct view of the source of light within an angle between 0° and 40° to the horizontal. For the purpose of avoiding this type of glare, projectors are normally used whose lamps are recessed into the casing or the help of accessories suitable for this purpose is employed.

Indirect glare is caused by the viewers' eyes being saturated by the reflection of light falling onto the exhibited work. This reflection of the light beam could arise either from the beam of light from a projector being used to illuminate the work or from patches of light from Up-lighters placed less than 60 cm from the ceiling, which, due to their great intensity, actually reflect uncontrollably on the work exhibited.

In the case of glare caused directly by the effect of a light source, normally projectors, there are different opinions amongst experts, some of whom say that the maximum angle between the illuminated vertical plane and the centre of the beam of light should be 30° at a distance of 1.05 m from the painting.

Others say 15° for the same angle, and a distance of 0.8 m, which gives different results for the distance of the projector from the wall. In these two examples, the difference would be 6 cm, taking the tangents into account, since the distance is the same as the height at which the projectors are set, less 1.6 m (the average height of the eyes of a Portuguese) multiplied by the tangent of the limit angle.

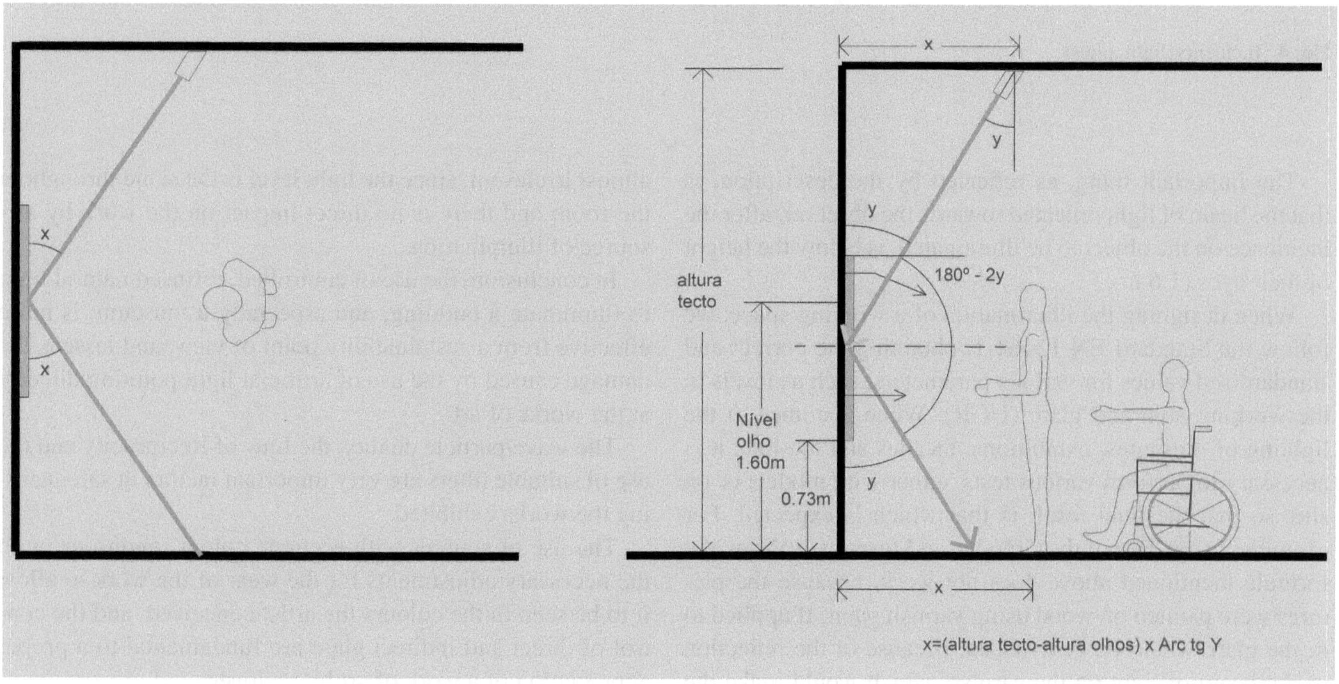


**Fig. 3** Light ergonomics

**Fig. 4** Incidence light angles

The important thing, as reflected by the description, is that the beam of light reflected towards the observer, after the incidence on the object to be illuminated, is below the height of their eyes (1.6 m).

When designing the illumination of a working space, we follow the Standard EN 12464-1, obtaining the correct and standardised values for various parameters, such as levels in the working area and glare (UGR). When it comes to the lighting of museums, exhibitions, facades and the like, it is necessary to perform various tests, either with models or on site, so that the final result is that which is expected. For example, in the case of the Grão Vasco Museum in Viseu, the formula mentioned above does not work, because the pictures were painted on wood using varnish paint. If applied to it, the glare would be generalised, because of the reflection of the beam of light on the varnish, which would make the reflection uncontrolled.

The secret, in these cases, is to illuminate the area around the picture, keeping the environment in which it is hanging at the same levels as those in the plane of the work. In the case of the use of a controlled, diffused natural light, as happens to some degree in the Grão Vasco Museum, the Serralves Museum and the Iberê Carmargo Museum, the glare is almost irrelevant, since the light level is the same throughout the room and there is no direct impact on the work by any source of illumination.

In conclusion, the use of controlled, diffused natural light to illuminate a building, and especially a museum, is more effective from a sustainability point of view, and lessens the damage caused by the use of artificial light pointing directly at the works of art.

The wave/particle duality, the Law of Reciprocity and the use of suitable filters are very important factors in safeguarding the works exhibited.

The use of sources with accurate colour renditions, with the necessary adjustments for the wear of the work to allow it to be seen in the colours the artist conceived, and the control of direct and indirect glare are fundamental to a proper appreciation of a work of art by art lovers and visitors.

Carrying out lighting tests using models and/or in the course of setting up and fine-tuning the assembly of the permanent exhibition are both crucial for the results to be pleasing to all. Without these considerations, however many theories there may be, it is not possible to reach a positive outcome that safeguards the integrity of the works exhibited and their potential for proper viewing by visitors.

# From the Utilisation of Rainwater to the Re-use of Grey Water: Some Thoughts Concerning the Iberê Camargo Foundation Museum

Armando Silva Afonso

## Introduction

The design for the Iberê Camargo Foundation building in Porto Alegre, which provides for the utilisation of rainwater and the re-use of grey water, raises some thoughts on these solutions and on the situation in Portugal with regard to their implementation.

Rainwater usage systems in buildings have undergone considerable development in various countries, particularly Brazil and Germany, not only for reasons of rational water use, but also as a process of reducing flooding peaks during periods of rainfall.

This usage has also been given increasing attention in Portugal [1], through a civil society association that promotes water quality and efficiency of water supply and drainage in building installations (ANQIP – National Association for Quality in Building Installations). In the absence of Portuguese Standards or European Directives, this association has developed a Technical Specification for the design, construction and maintenance of these facilities in buildings.

It should be pointed out that, from the perspective of the rational use of water, the so-called Mediterranean climate is apparently not conducive to the recovery of rainwater, given that it is characterised by hot, dry summers and cold, wet winters, and the dry spell in summer usually lasts for 2–3 months [2].

As the name itself implies, this kind of climate is only manifest in the Mediterranean basin, although similar conditions can occasionally be observed in the south of Australia and on the east coast of North and South America. Some European countries, like Portugal, Spain, Italy and Greece, are partly or wholly covered by this type of climate.

With regard to the re-use of grey water, the situation is similar in Brazil and Portugal, where there are no official standards as yet. In Portugal, ANQIP has also developed a Technical Specification, but in this case, the focus is placed on issues of quality versus usage.

In general, water efficiency measures, in addition to meeting an environmental imperative in any country in the world, are, in Portugal, a strategic necessity, given the risk of water stress.

In fact, Portugal has a high risk of hidric stress in the short/medium term (Fig. 1), so the utilisation of rainwater and the re-use of grey water in the context of promoting overall water efficiency in buildings can be an important, albeit insufficient, contribution to reducing this stress, and could also have other complementary beneficial effects, like the reduction of stormwater flood peaks [2].

## Water Efficiency in Buildings: The Principle of the 5Rs

The overall inefficiency in the use of water in Portugal is currently estimated at more than $3 \times 10^9$ m$^3$/year, or approximately 39% of the overall water demand in the country.

With specific regard to the urban water supply sector (public and residential systems), the total inefficiency is estimated at being close to 250 x $10^6$ m$^3$/year, representing an economic value of around 600 x $10^6$ €/year.

In terms of per capita indicators, reported values correspond to inefficiency greater than 25m$^3$/year per inhabitant, worth around 60€ per year per inhabitant. In view of the projected water stress in the short/medium term, this situation is unsustainable and requires immediate action, through the implementation of measures to rationalise water use.

Note that the rational use of water in buildings can be summarised by a principle similar to the 3R (usually applied to waste), but more comprehensive, known as the 5R principle (Fig. 2) [3].

A. Silva Afonso (✉)
Department of Civil Engineering, University of Aveiro/RISCO, Aveiro, Portugal
e-mail: silva.afonso@ua.pt; anqip@anqip.pt





B. Rangel et al. (eds.), *Museum Technology and Architecture*, Building Research: Design, Construction and Technologies,
https://doi.org/10.1007/978-3-319-76172-5_5

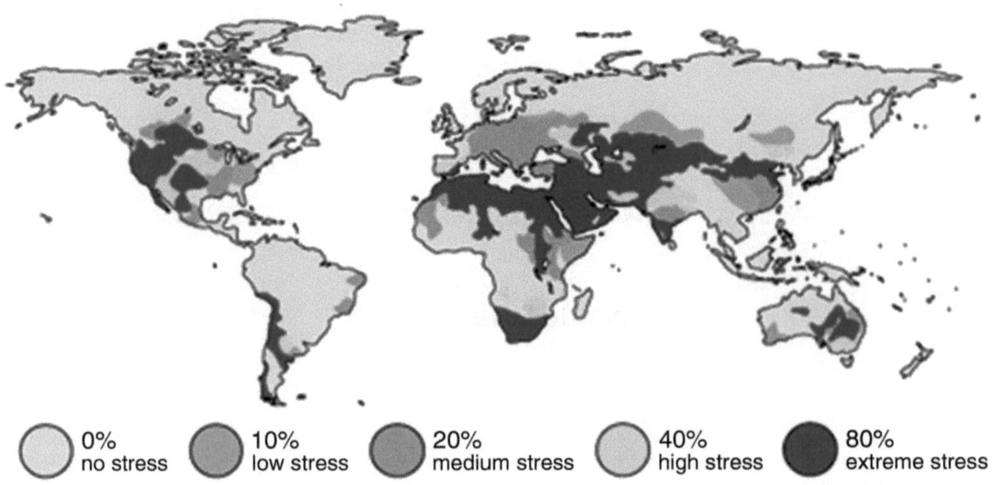


**Fig. 1** Hidric stress. Situation in 2025. (World Water Council)

- *Reduce consumption*
- *Reduce losses*
- *Re-use water*
- *Recycle water*
- *Resort to alternative sources*

} - *WATER EFFICIENCY IN BUILDINGS*

**Fig. 2** The principle of the 5Rs in water efficiency in buildings

The utilisation of rainwater falls into the category of the fifth R (resort to alternative sources) and, as mentioned above, has been the subject of the development of a Technical Specification by ANQIP (Specification ETA 0701) [4].

ANQIP has also developed a model for the certification and labelling of water-efficient products [3] and, as previously mentioned, has drawn up a technical specification for the re-use of grey water in buildings. It should be noted that the concept of recycling or re-use differs, in that water is treated and reintroduced into the building cycle or is used sequentially for different purposes in the building.

## The Portuguese Specification for the Utilisation of Rainwater in Buildings: Description and Comparative Analysis

### Introduction

Given that it is a specification from a non-Governmental body, Specification ETA 0701 compliance is naturally voluntary. This Specification is divided into 6 chapters (Introduction, Definitions, Statutory and regulatory references, Certification and general aspects, Technical requirements and Maintenance), and the Certification of these installations by ANQIP is recommended.

This recommendation is justified by public health and requires ANQIP's prior assessment of the project, inspection of the work and certification of the installers. For this purpose, ANQIP also developed a Technical Specification (ETA 0702) [5].

### Technical Requirements

For studies of rainfall, ANQIP mapped the average rainfall in Portugal, as shown in Fig. 3, based on information available at the Meteorological Institute. One aspect that was given particular attention was the need to divert the first flush, especially as prolonged droughts may accumulate pollutants on the roofs and aggravate the problems of pollution in this water. The installation of automatic diversion systems is recommended in the Specification.

ETA 0701 indicates criteria of time or amount of precipitation in determining the volume to divert. In the first case, it is thought that the initial 10 min of rainfall should be diverted, though it could be less (but never less than 2 min) when the interval between periods of rainfall does not exceed 4 days. As for the level of precipitation, the benchmark value is considered to be 2 mm, but this could vary between 0.5 mm and 8.5 mm, according to local conditions and the interval between rains. The Brazilian Standard also adopts 2 mm of rainfall as its criterion, while the German Standard, DIN 1989, has no requirements in this respect [4].

ETA 0701 also requires the fitting of appropriate filters in the connection to the reservoir (for retention of leaves, etc.), as in both the German and Brazilian Standards. The Specification also includes technical prescriptions for preventing contamination in the discharge of water from the overflow of the cistern, from the diverting system of the first flush and from the rainwater filter. The destination of these

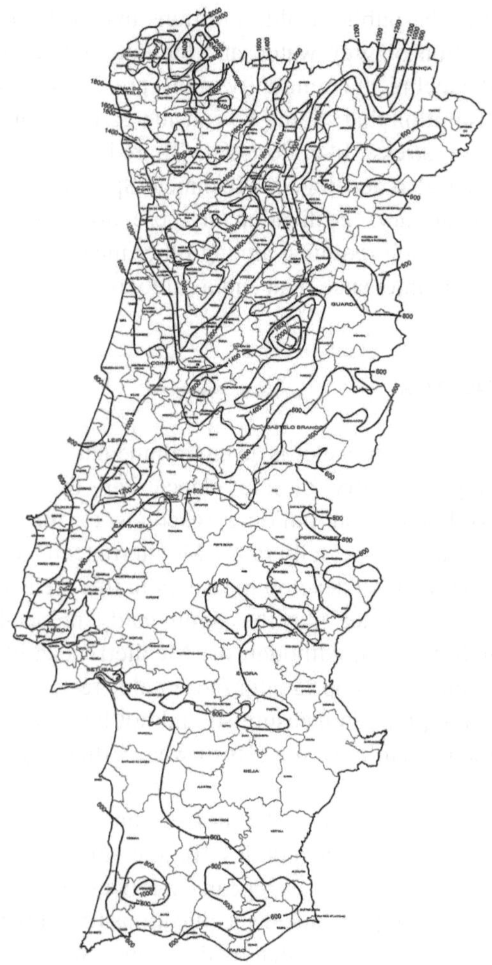

**Fig. 3** Map of average rainfall in mainland Portugal (ANQIP) [4]

waters may be infiltration, discharge into a natural water course or into a drain.

ETA 0701 also requires a device to be installed that reduces turbulence and decreases the velocity of entry of water into the reservoir. The suction pump should also ensure a low suction velocity, with the entrance between 10 cm and 15 cm below the level of the water in the cistern (or by an equivalent system that does not allow the suction of floating or sedimentary residues).

The Specification also includes various construction measures, recommending that rainwater be stored in a place sheltered from light and heat, and that openings to the outside be provided with anti-rodent and anti-mosquito devices. It is also necessary to install a cut-off at the beginning of the system, so that when substances potentially harmful to human health are used or spilt (deliberately or accidentally) in the catchment area, the system can be disconnected, preventing these substances from getting into the cistern [4].

The dimensions of the reservoir are one of the issues that have caused the most disagreement amongst specialists, and there are numerous applicable methods, from the simple models (abbreviated German model, simplified German model, simplified Spanish model, English practical model, Azevedo Netto model, etc.) to the theoretical and probabilistic models (Rippl model, simulation model, Monte Carlo model, etc.)

In the Portuguese specification, it is proposed, given the country's climate characteristics, that a sizing of the cistern based on a monthly balance between availabilities and needs be performed. In any case, the storage period is limited to 30 days in current situations, but can be increased by up to 90 days for some uses, such as watering. This maximum period is significantly longer than that in, for example, the German Standard (3 weeks), but was based on studies of water quality (physico-chemical and bacteriological) carried out by ANQIP in experimental facilities.

To facilitate the determination of water demands in the building, the specification includes a table of standard consumption per device installed, as in the German Standard. The Portuguese table is based on the use of devices labelled category A for water efficiency, according to the ANQIP certification scheme, since the use of a rainwater harvesting system with non-efficient devices is not considered logical. The German and Portuguese tables essentially differ with regard to the values for irrigation of outside spaces, which is understandable, given the different climates [4].

As in the foreign standards mentioned above, it is required that non-potable water networks, including accessory parts, be clearly differentiated from drinking water networks. Devices for watering or washing, inside or outside, should be labelled with identifying signage and warnings. It is also recommended that taps for washing or irrigation be provided with removable handles (security keys) to prevent unsuitable use.

The installation of a water meter should be considered in the section connecting the tank to the building network, in which the flow rates that are not channelled to the drainage network (garden irrigation, etc.) need not be measured. This requirement derives from the Portuguese legislation on water measurement.

Issues of quality give rise to significant divergences amongst the specialists. The utilisation of rainwater to wash clothes, for example, is not allowed in Brazil, but is permitted in Germany. This difference in criteria will result, eventually, from the different washing temperatures considered and their respective effects on micro-organisms.

The Portuguese specification is closer to the German Standard, considering the following potential uses:

- Flushing toilet bowls
- Washing clothes
- Washing floors, cars, etc.
- Watering green areas
- Industrial use (cooling towers, fire networks, HVAC, etc.)

The utilisation of untreated rainwater to flush toilets, however, may only be allowed when the water meets, at a minimum, the quality standards for bathing water, in accordance with the applicable European Directives (Council Directive no. 76/160/CEE, of 8/12). If necessary, disinfection should be considered [4].

With regard to washing clothes, using rainwater that has not been specifically treated should only be considered when the water temperature reaches at least 55 °C. A micro-filter with a mesh of at least 100 μm should be installed for this usage.

If the pH of the rainwater is usually less than 6.5, correction of the pH is considered to be either necessary or convenient, according to the materials used in the installation.

The supply system to the cistern from the drinking water system, for the purpose of preventing a lack of water in the cistern during periods of prolonged drought, should be made in such a way as to prevent contamination of the drinking water system. ETA 0701 also includes various notes and recommendations about the characteristics of the pumping equipment and its installation [4].

## System Maintenance

ANQIP Technical Specification ETA 0701 includes a table indicating the frequency of maintenance work. This table is similar to that in the Brazilian Standard and less exhaustive than the one in the German Standard [4].

## Brief Note on the Re-use and Recycling of Grey Water

As mentioned above, ANQIP has also drawn up a Technical Specification for the re-use of grey water in buildings (ETA 0905) [6].

In some countries, like the USA, the present tendency is to use equipment already available on the market, at least with regard to existing installations, and it is noted that the applications proposed for the re-use of (untreated) grey water are essentially for sub-surface irrigation.

The use of gray water in toilet discharges, without treatment, is possible with reduced retention time, using compact equipment that is available on the market, such as lavatories combined with a toilet. In public or collective buildings with high periods of grey water retention, more or less developed treatments should be considered for permitted uses (toilet discharge, washing floors and surfaces, etc.), which may include, in the most complete systems, biological treatment, filtration and ultraviolet disinfection.

Despite the public health questions that still arise about the utilisation of grey water, and the doubts that still exist within the technical community in this area, it is evident that this is a solution that should be considered in countries with high water stress, like Portugal. To reduce health risks, ANQIP has also developed a Technical Specification for the technical-sanitary certification of these installations (ETA 0906) [7], which requires, among other measures, the existence of a maintenance plan and a safety plan [8, 9].

## Conclusions

The efficient use of water is an environmental imperative for every single country on the planet, and the new Iberê Camargo Foundation building responds to this concern. Some countries, especially in the Mediterranean basin, must take urgent steps to develop measures in this area, since the availability of water may be significantly reduced in the short/medium term. Although the Mediterranean climate is not particularly appropriate for the suitable re-use of rainwater, for example, this measure should not be ignored, being one of the principles of the 5Rs of water efficiency in buildings [2].

This is why ANQIP, a Portuguese non-profit association of companies and universities, in the absence of either Portuguese standards or European Directives in this area, has decided to develop technical specifications for rainwater harvesting in buildings similar to those that already exist in other countries, such as Brazil and Germany, as well as technical specifications for the re-use of grey water and for the sanitary certification of these installations.

The adaptation of Technical Specification ETA 0701 to buildings with green roofs, another solution of major importance for the environmental sustainability of cities, is currently being studied by ANQIP [10].

## References

1. Castro R, Silva-Afonso A (2007) Integration of sustainability in sanitary installations: the example of the Aveiro DOMUS House of the Future. In: Proceedings – SB07 sustainable construction – materials and practices, vol. 2. Lisbon, Portugal
2. Silva-Afonso A (2009) Building rainwater harvesting systems. Doubts and certainties. In: Proceedings CIB W062 2009 – water supply and drainage for buildings. Dusseldorf, Germany
3. Silva-Afonso A, Pimentel-Rodrigues C (2008) Water efficiency of products and buildings: the implementation of certification and labelling measures in Portugal. In: Proceedings – CIB W062 2009 – 34th international symposium on water supply and drainage for buildings. HKPU, Hong-Kong
4. ANQIP (Nacional Association for Quality in Building Installations) (2015) Technical specification ETA 0701 – rainwater harvesting systems in buildings (version 9). ANQIP, Aveiro

5. ANQIP (Nacional Association for Quality in Building Installations) (2011) Technical specification ETA 0702 – rainwater harvesting systems in buildings (version 3). ANQIP, Aveiro
6. ANQIP (Nacional Association for Quality in Building Installations) (2011) Technical specification ETA 0905 – rainwater harvesting systems in buildings (version 3). ANQIP, Aveiro
7. ANQIP (Nacional Association for Quality in Building Installations) (2011) Technical specification ETA 00906 – rainwater harvesting systems in buildings (version 0). ANQIP, Aveiro
8. Lança I, Silva-Afonso A (2008) As Alterações Climáticas, as Medidas de Eficiência Energética e a Saúde Pública. Uma Análise ao Nível das Instalações Prediais. In: Proceedings – XIII SILUBESA – Simpósio Luso-Brasileiro de Engenharia Sanitária e Ambiental. ABES, Belém do Pará
9. Silva-Afonso A, Pimentel-Rodrigues C, Lança I (2011) Re-use and recycling of grey water in buildings. The Portuguese approach. In: Proceedings CIB W062 2011 – water supply and drainage for buildings. Aveiro, Portugal
10. Silva-Afonso A, Pimentel-Rodrigues C, Tadeu A, Almeida R, Simões N (2016) Rainwater harvesting in buildings with green roofs: runoff coefficients. In: Proceedings CIB W062 2016 – water supply and drainage for buildings. Kosice, Slovakia

# Part II

# The Coach Museum_Lisbon

Project by:
Paulo Mendes da Rocha
Ricardo Back Gordon
AFA consult

Coches Museum
Street view from Av. da Índia
Copyright Arménio Teixiera

# Display of Technique in Paulo Mendes Rocha Works

Bárbara Rangel, José Manuel Amorim Faria, and João Pedro Poças Martins

BR – Do you see architecture as a meta-discipline that seeks to bring together interdisciplinary knowledge?

PMR – Yes, I do absolutely consider it a meta-discipline. We are working with the imagination…

Nowadays, global awareness has given architecture an extraordinary level of importance in the context of knowledge within universities, as it calls upon knowledge of the issue of construction of the contemporary city, which is, after all, the habitat of man. Architecture doesn't stay still, it moves ahead with the concept of a peculiar form of knowledge, specific to the human race itself. Awareness of the building of what one imagined, in other words, everything we do was there before in the imagination. The only way of existing before in the imagination is, in a constructive manner, combined with all the other ingredients of desire. It's a goal, a form of knowledge.

JAF – In this sense, what is the importance of technology in architecture?

PMR – It is almost as indispensable as in language, in the forms of building. Amongst ourselves, we say, it's not just me: 'Words are to a poet what stones of a cathedral are to us': it's a construction. It's a beautiful image that perhaps they created at my school.

Imagine, words melted into me. Whole. Words. If you break this verse down for someone who doesn't know it, they'll say: What's this? That's nothing, you take one word here, one there, another there…

Everything for us is construction: life itself, home life, the measures you take to exist. In my opinion, it's a vision of architecture. From a purely practical point of view, you can copy, adopt rules and methods, and establish many ideas about culture and groups, and keep replicating them.

To me, this is a matter of flags, and they're not the most interesting forms.

I don't see much interest, particularly today, in distinguishing between a Brazilian architect and a Hungarian architect.

BR – Do you see technology more as a language or as a working method?

PMR – Technology is absolutely essential. It's as much one thing as the other. I'll tell you a story: Here in Brazil, there was a famous anthropologist and scientist, who was Minister of Culture in Brasilia at an interesting time before that awful military coup by Darcy Ribeiro.

At the newly opened University of Brasilia, we decided to hold an exhibition of construction technologies. The prominence of prefabrication was in vogue, or at least under discussion, particularly in concrete (which was not very common), to address problems of quantity, because there was a shortage of houses. They decided to do an exhibition in the city on prefabrication techniques, on-site demonstrations. When they went to speak with the Minister of Culture during the preparation of the exhibition, he said that not only did he support it, but he would also provide one of the exhibits. He simply did this: using a means he had of approaching the indigenous population, he called on a group of Indians (who still exist there) to come on foot, alone, and in every place they passed, to choose the timber to build one of their peculiar houses. The tribe lives in just one house. It's a circular dwelling 50 m in diameter, for 100-200 people. It's a hollow that looks like a sea urchin.

It's open in the centre. They live in a ring around it that's open because of the fire, a big fire they keep in the centre. It's open air and there's a covered strip of 8–10 m around it. The way they do this is perfect!

The woods are, in a sense, pre-existing, not prefabricated, but they're not standard. What do they do? They fix the timber, the poles, then they bend them with a length of vine nailed to the ground. With the torque, which can be regulated, it becomes like a circle, a perfect circular ring. Inside, it's slung with nets and they live in a collective manner.

B. Rangel (✉) · J. M. Amorim Faria · J. P. Poças Martins
Universidade do Porto, Porto, Portugal
e-mail: brangel@fe.up.pt



B. Rangel et al. (eds.), *Museum Technology and Architecture*, Building Research: Design, Construction and Technologies,
https://doi.org/10.1007/978-3-319-76172-5_6

The collaboration of Darcy Ribeiro in that technology exhibition led to the display of an extraordinary event! That group erected the entire hollow in 2 weeks, with resources that were lying around, after walking miles in the midst of the mountains. It could have been a bluff, it could have been timber from a truck!

In any event, the scenario was very interesting. Today we would say that it's pre-tensioned, both the beams and the cable. It's very beautiful!

RF – It has all been invented already.

PMR – It has, because – to support what Rui said – if necessary, we invent what has not yet been invented. The issue is that one of the major ingredients of technology and invention is urgency.

JAF – I want to ask a question about technology. The School of São Paulo, Philosophy, The School of Architecture linked to Philosophy. How do you see this connection between architecture and other groups of sciences, such as philosophy and thought in general? Do you consider thinking to be critical to guiding architectural creation?

PMR – I think it's vital, absolutely indispensable, it's the beginning. You invoked an occasionally recurring image, this story of the Paulista school (many get confused, mainly ill-advised critics, thinking it's a particular style, the exposed concrete, the minimalism, all nonsense…). If you can't find a pigeon hole for it, people feel lost! They don't have the confidence that what is "lost will be found." No! They have to find it immediately, and it ruins everything…

One distinguished professor, Vilanova Artigas, was fond of saying: "I am sure that you cannot teach architecture, but you can educate an architect."

That's the key to what we call the Paulista School, in my opinion. First of all, you need to understand that there's nothing new. It's not a school that initiates anything, but it does invite awareness about this issue, which then becomes fundamental indeed. It's like saying that all the other schools are also like this, but perhaps without the sharpness and energy! The autonomy of the School of Architecture was the opportunity it represented and the favourable space it occupied, because it was simply a course at the Polytechnic School and the Faculty of Philosophy, Science and Letters at the University of São Paulo. It had already been inaugurated with this spirit of dependence between critique and technique.

For us, it's like saying, in opposition, that there is plenty of technique that is not very good.

BR – In Portugal, we say that too much of a thing is good for nothing.

PMR – A certain hunger, a certain lack of food is part of high cuisine. You eat raw, as someone said.

BR – Do you see this allure of technology as a modernist vision or a classical methodology?

PMR – It is eternal to human consciousness! It's what we call a timeless thing, like, for example, our origin. The thing was ensured by the coincidence between the formation of consciousness and language, the sense of urgency!

BR – But this allure that comes from technology, present in your work, is it a modernist vision or a methodology? Just to provoke you a little, does it come more from your engineer father or from Brazilian modernists such as Vilanova Artigas?

PMR – I don't distinguish between the two, that's the thing. With all due respect and affection imaginable for my father, because I liked him a lot. I see the issues of architecture as higher engineering designs.

I better understood who my father was when I met Artigas, actually. I met Artigas quite late, he wasn't my teacher, since I graduated from another school. I was invited by him, with great ceremony, to be his assistant after that Paulistano competition. It didn't start out as a remarkable work (it's not just me saying that). It became remarkable precisely for the constructive ingenuity, on which its actual realisation depended considerably, by an engineer, Tulio Stucky, who was involved with the construction company that built it, whereas we developed the idea, we applied the idea of suspension cables, etc. I was lucky. The circle as an undeformable figure, once subjected to homogeneous forces. Very interesting! And why? I didn't end up with a technology I'd chosen before. I arrived at the concept of a large span as a space needed for this interplay.

It could be achieved with a suspended structure, with the great contradiction of a lesser height in the centre where it rests. I imagined suspending a ring, etc.

The technology that we ended up adopting was prompted by what we knew about doing what I wanted, which was merely not to do it in heavy concrete, in a large span. Maybe because we did not know very well how to do it.

BR – So, it seems to me that you follow a quasi-scientific methodology to address the problems. Is that your way of approaching problem-solving in architecture?

PMR – The idea of science is behind everything: tension, gravity, stability of construction, soil mechanics, fluid mechanics… The question of architecture as something solid, maybe related to my training in this connection with engineering that you have already identified, contains intriguing data. A child doesn't know it yet, but is delighted to see it. My father's specialty was inland waterways, harbours, rivers and canals, in other words, water. This management of the mechanics of the universe through the fluidity of water, which seems impossible, is what made me develop an open view on the question of stability of buildings, although everything is unstable. It's our position really: lost on this little rock. Abandoned in the universe. Abandoned, ma non troppo, as the Italians say, subject to the laws of mechanics. This is the key to knowledge of any constructive vision, which can even apply to words. The lexicon is nothing without the syntax.

BR – Incidentally, I was reading the texts of several interviews that you gave, and you often spoke about one of Vilanova Artigas's principles: "if you form a question really well, you already have 90% of the job done."

PMR – It's true! I would even like to emphasise it to make Artigas teeter on the brink! That's where the work begins. A job worth doing. It's no longer a subjective effort, it's an objective effort.

JAF – Turning now to the Coach Museum…

PMR – I found an extraordinary engineer in Sao Paulo, Sigero Mitsutami (of Japanese descent), educated at the Polytechnic, with whom I did much of the subsequent work. Not knowing about this extraordinary invitation to do the Coach Museum, but knowing how interesting this project was, I thought that I would only be able to do the project if I could count on the support of this engineer that I met on a visit I made to Portugal.

I made two interesting visits here, and discovered Rui Furtado. On the first, I found him personally at the stadium in Braga. I was taken there by a friend from Portugal, just like that, on the eve of the inauguration, without telling anyone. When I got there, Eduardo [Souto de Moura] and Rui were getting things ready for the opening, making sidewalks and completing the work. They caught up with me and walked me around the whole work. They showed me the spiking in the rock. It reminded me a lot of the Indians.

I talked a lot with him. I was delighted, and often, when I feel delighted with things, I don't stop talking. We decided to remain friends forever, but we rarely see each other.

On another occasion – it's easy to imagine the passage of time in episodes – someone took me to see Casa da Música in Porto. Rui was not there, but they told me that it had been him doing the engineering work, which wasn't yet completed, but was close.

Then, on that occasion, in my eagerness to eventually accept the invitation, I called Rui, telling him about my circumstances, to see if he would join me eventually. He immediately said yes. I said: it's all done. And on this occasion, I was delighted.

Something I already knew came up again, something that is very interesting in relation to architecture: while you may not know exactly what you have to do, you can be sure of what you should not do. It's somewhat linked to the idea of knowing how to formulate the questions.

The road is open, but is not resolved: you can get lost along the way that you yourself have opened. So, yes, there's a very interesting question of collaboration: to pursue precisely the good things along that way, because it's very easy. It's the image of a course made of already discovered, existing, diverging paths. There's an anti-beauty – I don't know if you have this word – it's a strategic path that someone invents to get there.

I saw clearly what we did not want to do there: the big question was what size should this necessarily large building be? The size of things is very interesting in architecture, and the numbers are a bit abstract, so many square metres, but who might have a view of the size of things? And this huge building seemed like it would destroy the place.

The first issue that you see is that this whole enclosure has an incredible museum value. There is much talk of the presence of Portugal in the wisdom of voyages, all that. This whole place is a museum of Portuguese sea exploits. The caravels left from here.

I think Brazil and Portugal, in particular, are very fond of it. It wasn't America that was discovered, but Brazil, as if it was already there. It is an anticipated dream, this idea of place, especially when considered from a healthier point of view than that of strictly business. Tourism today is very interesting, because the world is really beginning to see itself. Like the Indian who likes to put one hand over another's, otherwise he won't believe it. It's the whole place!

This building, so portentous, solid, broad, like a museum has to be to house a collection of huge artefacts, is so inconvenient in that place.

The beginning of an idea arose! The idea of lifting the building off of the ground seems a bit absurd at first… I'm telling you things that I put to myself: "But how? That thing, heavy, so big, you want to lift it off the ground?" Then, confidence in the technique appeared, because this sort of thing has actually been done quite a bit. Then, right next door, is the famous bridge, an example of the very best technique. If you had to name the most beautiful work in the world, it doesn't exist: what is the most beautiful? If you chose those we already know, the Taj Mahal would be there. But the Golden Gate Bridge undoubtedly stands a good chance. The structure exists, the structure is right there at hand.

At first glance, it would be outrageous to suggest lifting it off the ground – but why not? If you can lift a train off the ground, you can also do it with coaches!

Another very interesting thing that adds to the museology of the place that will be the museum: The museum idea is to make others see. To see something means to think about it.

Here comes the construction of the city in that place, very obviously from the museum point of view. Land reclaimed from the sea, confronting the fluidity of water and what is solid, and so on. Even my vision – which is not only mine, for sure, but which holds a lot of attraction for me – of this Lisbon from one side to the other of the Tagus, with the lights on, everything illuminated…

I see a transverse axis across the city, but in this particular place, already considering the nature of fluvial navigation, of passengers from one side to the other.

All this is very interesting to consider, all the more so because it already is there, in the commission, the brief, the pedestrian crossing to the railway track and the high-speed road – which already exists there, in a somewhat preliminary way – but it would be good to take it to the sea, to link it to navigation, since there is a small harbour there.

This association with the footpath, with the whole Jerónimos area, the old town, the small alleys around Rua de Junqueira, the names of famous streets, a little forgotten by the actual residents of the city. The small alleyway of the old customs dock, everything already inside a territory that was not really a territory[?]. The waters came up to there. Hence, this Portugal, a seafaring and solid country…

In the actual construction, the water table is very high, it is difficult, you're dealing with exactly the hardest thing. Raising it off the ground means concentrating the loads upon those few pillars. It was assumed that there would be few, as few as possible. I myself built a little chapel to St. Peter [chapel of St. Peter the Apostle], which they commissioned in São Paulo, in which, I don't know if it was for St. Peter, I put just one concrete pillar. The value of doing that!…

Lisbon is a city subject to seismic problems, which makes it even more difficult to suspend heavy things off of the ground with so few pillars.

But displaying the achievement of this difficulty is another museum piece, displayed as a virtue, not as perfumery. Because then, you can do the opposite.

In my view, I don't much like architecture that shows itself off, as if it was merely done on a whim. I'm introducing a need to free the footpath from the ground so that the small alleys are seen, as was the place itself.

Even here, between Jerónimos and the Coach Museum, there is an area with a small development with an almost… it reminds me of the Marais in Paris – Poultry street, Blacksmith's street… It's all "tumbled down" there… those alleys, you can't move – and I thought it would be a territory that required a delicate touch.

JAF – And the issue of siting? Why are these axes there?

PMR – Yes, this attraction of the place we're trying to explain, or praise, from the point of view of the museum, the axis of the slope, the Calçada da Ajuda, covered with monuments, perpendicular to those waters we were talking about, is very important. And the land presented a deformation, on the corner of Ajuda and Junqueira, in relation to the "tumbled down" part along the plot, that stretches along the front of the Tagus, which is 300 m of land. The building itself is 150 m. It has to be big to fit the coaches in.

That just made me think about the technical issues and the demands of technology. The architectural problem now: how to transform large exhibition spaces inside a floor layout? In this case, it's a more emphatic museum than any other, because there are many objects and they are very large.

How do you make a plan of a house? The corridor, bedrooms, etc. How do I move from the great halls to the offices, the study rooms, between the small and the large, as halls, enclosed spaces?

I remembered that it would be interesting to do an annex to address these issues in this enclave of land near the corner of Junqueira and Ajuda, which immediately gave it a special charm – all of this took place in my mind. Because to link the two buildings, I made a small bridge and enclosed an internal "piacetta" – as the Italians say – at the bottom of the "tumbled down" houses.

The whole project was done in the mind and in conjunction with Rui. That's why I remembered the engineers, my former comrades from construction, and we began to talk together, so to speak. As if one were seducing the other: "I know what you want, it's this and that". And it was.

BR – Speaking of this moment of seduction, what was the magical moment when you found and generated the question?

PMR – The first time we met, there in the Souto de Moura stadium. That was the magical moment. I knew the people and I already knew we were going to do something together!

BR – Was Rui with you when you came here to decide whether or not to do the job?

PMR – No, I came before. I would never put a friend in a bad situation…

It was my responsibility to see to what extent we could do something. Then I said: "Rui, I have a very interesting project for us to do."

Then, other colleagues came on board. Bak Gordon (I thought I would need his support here). There is a practical issue in regard to this project involving heritage, the zeal that Lisbon and Portugal have for this place. He needed support with regard to the laws and applied technology standards for the place. Attention and dedication to the work in situ. So, I invited my colleague, Ricardo Bak Gordon, who accepted, and also Nuno Sampaio, for the questions of museology. We set up a working group and started to develop the project.

The metal structure technique we used is more or less down to a matter of engineering, of ingenuity. The metallic structure allows for easier prefabrication. On this fragile ground, the profiles are industrial, it all seemed very interesting, hence the construction in metal.

JAF – And the structure is lighter, and thus enables the transparency of its architecture.

RF – And especially, given the seismic issues in Lisbon, any solution in concrete would be disadvantageous from the point of view of foundations.

PMR – A metallic structure is always more flexible! Nothing can be rigid in architecture.

If it weren't for cloth, rope, there'd be no ships. Everything must be flexible and articulated, or else it all breaks up.

RF – Just yesterday, we were there on site to look at how to resolve the issue of the walls. That light solution of interior and exterior panelled walls.

PMR – Yes, this building has to be air conditioned, so there's nothing better than a thick metal structure.

JAF – And with little weight.

BR – Let's go back a little bit to understand how the technology started to become integrated into the team's creative process. In this programme, you mentioned that there is a dense, closed object that is the museum and an open, sweeping object that is the annex – what is the technological response within each of the buildings? I ask this, quoting a phrase in your last lecture that I heard when Rui was with you on stage: You said that with the help of Rui, you would indulge in any folly. What was the folly?

RF – With the help of Rui you'd indulge in any folly!

JAF – It's provocation…

PMR – It can happen in a jazz group. Any drummer can say: "With Miles Davis I can indulge in any folly…"! You have to go full circle, and instead of fixing it up with the drum kit, you do it with the saxophone.

You can turn a folly into something very interesting. They're variations on a theme. Music is a very interesting thing to consider within the context of architecture. I don't see the resolution as being musical, that's not the idea, but it is how we work.

I really like the limited amount of resources we have: it never hurt Rachmaninov that there were only seven musical notes!

BR – This is a question for both of you: Rui, what was the folly in this case, and what was the technological challenge of a building that is closed, opaque and heavy and the annex that is a sweeping building in concrete? They're really two very different technologies…

PMR – Here, in the main – or rather the larger, building, because there's no main building any more – another interesting issue in architecture is that it is not made of parts that can then be separated. What was apparently done in parts then becomes an integral whole.

Even the annex, with its more concrete appearance, is the same as the other. It has a concrete structure that supports metal constructions. There is a metal roof with glass and a 50-metre transverse, made with four main steel beams with the same design from the point of view of technique.

This land facing the sea – as in all of Lisbon facing the Tagus – is unsuitable for underground work, or it's only achieved at the cost of employing large-scale construction technology. So, we thought we'd do a car park for the area and not just for the museum, which is nothing new. The ingenuity of this continuous circular parking ramp already exists.

We proposed a parking lot next to the pier, linked to the pedestrian footpath that starts inside the Annex. We did this so that the footpath would not, as is common, be seen solely as a secondary element: it has become very important.

The question that arose was one of access for people, because the entire structure was now tied to the transverse, so as to accomplish the halls.

We made a building of 150 m, with two halls. A question of architecture, of strict functionality, arises here: the size of this structure, with such large coaches. It seemed that the hall could have a transverse. We considered the possibility of constructing it in steel with the need for space. 20 m is fine, it's manageable: so there are two halls of 20 × 150, because the transverse is possible.

All of this gradually started to lend coherence to the project. However, this systematic transverse structure is not amenable to being interrupted for the purpose of making ramps and that sort of thing. So we used the lift for access, which, from the museum point of view, goes far beyond the concept of the Coach Museum: this laying out of the city according to horizontal movement – whether by virtue of the ships primarily or now the trains – and vertical movement, which is given by the lifts. Without the lift, there is no contemporary city. If I had to say what the main ingredient of the contemporary city is, it would be the elevator. It allows for concentration, tall buildings, etc.

It's an interesting question of architecture, once again, because it's not enough to make a building with 100 storeys! It seems silly! No one lives in such a place; domestic life doesn't happen at a distance from the ground of over 60 or 70 m. It's foolish. For now, it's nonsense. You're abandoned inside your own exclusive house. These issues are not mathematical…

RF – Actually, the photograph of this scale model reveals that, in fact, there was no folly, because Paulo has a very strong knowledge of structure.

BR – Serious folly.

RF – Folly in the sense of taking a serious risk. Paulo conceived of the original idea for the project. It's what is here, in this picture, this photo, which corresponds exactly to what is there, nothing more, nothing less, everything, including this outside walkway. Everything!

This is the idea that Paulo brought to present to the minister when he decided to accept the project.

BR – You had already discussed it?

RF – No. We had spoken very generally about it. We went to present the project, the general idea of the project, but when Paulo explained it, I listened and said, "Right, it's good." From there …

BR – 90% was done?

RF – 90% was done. What I would say is that the solution itself contained whatever was necessary to accomplish it, which is the main objective of an architectural project when it is born. You then have something to work with.

Of course, then – as we were saying earlier – the construction systems changed, the very constitution of the walls, the need for earthquake protection that Paulo wasn't so used to in Brazil, all of this was introduced later, but again, everything was there, because the solution was perfectly foolproof against any unpredictability or folly.

The follies that Paulo refers to, jokingly, had to do with the later development of the details of the project and with the exceptional discussions and great deal of fun we had in carrying out various things.

Then, of course, with regard to details, the approaches of the architects are very different amongst themselves. Things that for me – according to other views – wouldn't be very interesting were very interesting for Paulo, because one of the objectives of all of this was really to show the success of the technique and show it in a very clear way. For example, I was trying to hide the supporting equipment in the project. When he saw that, Paulo said, "No, no, let's put them in! I want to see that," because of his experience in Osaka.

So, all of that learning how to see the building was really interesting. But, here, it would be me, of course, that would indulge in the folly!

PRM – So, the museum goes beyond an idea of coaches as simply artefacts. The whole history of transport systems, vehicles, is more or less on display, including this aspect that Rui mentioned.

BR – And the demand for rigour, showing the building to the city and showing the city to those inside was, it seems to me, the reference.

PMR – This axis is very evident! The ramps within the building become evident. You enter the Coach Museum and cross the bridge outside and things start to appear that- I don't know if that was the architect's intuition, or if it was luck…

The level required for this pedestrian passage out here, necessary because it has to cross over the railway, coincides exactly with the elevation of the hall. So, people inside the museum only have a small slit – like an Eskimo's eye – in the two heads, to see someone out there at the same level: one is inside the museum, another is outside. All this has a charm that can't be announced in advance, but we presume that visitors will be able to enjoy this new situation a little.

An interesting situation that arises, for example, is that, as the ceiling height inside the halls for these large exhibits (every museum should have a good ceiling height, but in this case, at least 10, 8 m) means that the administration that we put outside here can use a small bridge that enters the museum, we left a 10-metre gap to make a portal for that internal piacetta through which you pass to get to the administrative office by an aerial passageway inside the museum. This is very interesting for special, distinguished, visitors, whom the management wishes to give a quick tour of the museum without interfering in the lives of those thousands of visitors who are there at the same time. And it is also useful for daily maintenance, which is performed by personnel from the workshop downstairs.

In the small pavilion, with 10 m now, the same 10-metre ceiling span, inside the museum, you see that every passage is presumed to go from one hall to another. I've already said that this structure is metallic because the diagonals allow this passage.

Then, there's a small pavilion for the public lifts and one on the ground floor, as fluid as the place already was, because on the day we visited for the first time, I had a coffee on the sidewalk at this corner.

JAF – The famous Belém cakes…

I want to go back to details. All museums also have a functional part, and this one even more so, because the coaches require a lot of maintenance and they're heavy vehicles, so they need to be taken care of in situ. One of the things I find remarkable about this project is the ability to reconcile a coach maintenance factory with an itinerary for visitors.

How did the idea of the lift and the functional aspect of the coaches arise?

PMR – These floor workshops are shielded, but they have large windows. The public can watch the work going on there from the floor, etc.

There's also a cafeteria sheltered by this overhang[?], here in the square, and there's a small pavilion we built in glass, to be transparent, where the elevators and the shop are. It's a pleasant, fluid floor.

JAF – The transparency really combines the two things: the visitor and the maintenance work.

And, by the way, lifting the coaches is also a brilliant idea, the way the coaches go up to the next floor up.

PMR – It has to be a machine! After all, we're in a port area. Moving stuff around can't be difficult for those who work in the port. What I'd like is for any child to be able to understand all of it easily.

What I like, in the face of the great deal of explanation that happens with any project, is that it seems there's a strict reason for things. We're not really interested in making architecture that, from the point of view of functions, ends up entrapping people. What architecture is all about is helping with the unpredictability of life. That's really an amusing place, so much so that we're imagining that these little houses with their backs facing here will gradually turn into bookshops, restaurants, etc. This will cause the place to be put to use. I hope. Because the backs are facing this piacetta, like the restaurant with outdoor tables, steps going down from here, well, we landscaped it. It's the assumption. Other things can be done, of course, but it seems fine like this.

RF – I'm not going to talk about engineering, but rather architecture. It's true that anyone who looks at this scheme, which is the project's initial scheme, understands that it's all exactly in there. That is, from the functional point of view, the museum is like clockwork and achieves all of those cross-flows and functional crossings that are required. From the constructive point of view, it's all here as well. Which means that this here – for which Paulo says there isn't a reason or that the only reason comes from an initial intuition – in fact, encompasses everything that the architectural design is. It's all here. Everything! It's impressive and very interesting, because of Paulo's work method.

Another thing that I think is important – that Paulo mentioned several times and that was a novelty for us, because that's not really the Portuguese school – is this question of permanently taking advantage of technique to show it as a part of the museum itself.

It's to show the building and use it as an exhibit in and of itself, that is, to explain to people how the building works and the reason for each of the things inside it. This was very interesting for us, because, on the one hand, it makes us more responsible for the design, but on the other, we are showing people what we're doing…

PMR – The directors cross this bridge from up there, here, the buses enter. There must be a widening in relation to the bridge itself. It's because from here onwards, it's shielded by the glass inside the bridge. And the museum's entire security system is in these 10 m. This means that the museum opens up in a discreet way. I thought of these things to myself, to have fun: Imagine someone was there with a bunch of keys, with lots of people, asking permission to open the museum! Here, security personnel enter to man every door in the museum. And through this bridge here, they're inside, and you see that the museum opens up from the inside to the outside. Discreetly, like a private club. In other words, our behaviour is really dictated by very simple motivations…

BR – In the sense that Rui was talking about, the original idea for this project arose out of a very inclusive perspective in regard to the building. From both the technological and urban points of view, and consequently, as a whole, the architectural point of view as well.

PMR – Rui Furtado never told me it was difficult. On the contrary, he said, "Let's do it like this," including showing many of the seismic system joints, and all that, in the building.

These are very interesting lessons to consider, lessons that the building should teach[?]. You see it straight away, if you pay attention! When I say you, I'm thinking of the young, the ones who are coming on the scene now. The idea that moves us is education and teaching. As a philosopher whom I admire says: we know that we were not born to die, we were born to continue. So, the possibility of seeing all of that is very interesting.

If the museum could be visited by a teacher and his students, making this speech about earthquakes, navigation, coaches driving past, you don't even need to enter the museum; you could leave it until later… The pretext is the coaches, which are early forms of transport. We use projections a lot, to show their history.

BR – This inclusive view we've been talking about is quite visible in the paper models that you make in your most intimate moments, when you're alone.

PMR – They've even made a book [of paper models], but it's not about me making paper models, it would be quite difficult to do it any other way. With paper, I mean, scissors, wood, simple things lying on the table.

Because, in that respect, all architects make models, it's not just me. But I enjoy making models very much, instead of scribbles on paper. It's easy to cut paper, pasteboard, cardboard, because you get a critical view of each version[?] at the same time, you're showing it to yourself, you question yourself!

RF – And it's the right time to make them, I admit, but I can imagine that the right time to make them gradually improves our reasoning.

BR – In the same way that Siza uses his notebook and biro to discover the spaces, you use scissors and paper, which is easier…

PMR – Perhaps I don't have as fertile an imagination as Siza, because I can't understand my scribbles as well as my models. They're more elementary, simpler.

BR – And from this model, how did you move on to the structure?

PMR – That's the engineer's job, that's where Rui comes in.

RF – For us, this is very clear, from any engineer's point of view…

PMR – It's the coaches, the objects we're displaying, that require the great ceiling height. If I'm turning the wall into the beam, why not? This means that the walls must be 10 m high, at least. The beam will require a height (the H in the equation) that will allow it to suspend the building from the floor… But I have 10 m and I'm using it all as a beam, a steel beam now, it's obvious that a 10-metre high beam withstands any type of span (50, 60, 70 m).

If I have a 150-metre building, the arithmetic is easy: 25 m cantilever, 50 m span, 50 m span, 25 m cantilever, with 3 columns, it's sorted out!

One interesting thing we haven't mentioned, but that is fundamental in architecture, is something that is banal but fatal: human scale. If you get close to a building where one column is 50 m away from the other, you end up seeing no column at all!

It's very interesting to show a quality that arises from the thing itself, like the Indian hollow. The ingenuity of the construction is virtuous in itself.

RF – From the design point of view, the level of integration you're talking about is that, because you don't just see the part here, you also don't just see the solution of the whole functional issue. You also see the poetic component of the construction that results from the articulation of the rest. That's the real integrated concept for the whole project.

BR – Almost as if the construction or the technology elicits the intended emotions when you visit the building.

PMR – The example – with which you may foster anyone's imagination - is the Gothic stone cathedrals, because all of the shapes there depend on a strict geometry in stone cutting, something that doesn't seem obvious. The charm of engineering is precisely the ability to say that we want to surprise nature and make it reveal its mysteries and transform its challenges into qualities. So, if stone falls, the way I cut it prevents it from falling again, and I make an arch.

So, it's pure geometry, as simple as it comes. Like the pyramids in Cairo, the great beauty that makes them stand forever, their wonder and charm is that they are still the machines of their own construction. It's the inclined plane, otherwise the stone cannot be carried to the top! And from this point of view, everything seems very easy.

JAF – There was mention earlier of the issue of transparency, linking the walkway to the inside. This question is twofold: first, illumination, because there's an issue here of natural light versus artificial light. And second, concerning the trades, from the perspective that the museum makes great demands in terms of comfort. It has heavy-duty air conditioning systems, which needed to be housed. According to the thread of our conversation, that is, the model that "contains everything," I'd like you to explain how you solved those two problems.

RF – Illumination is an issue that doesn't arise in a museum – and specifically in this one – in terms of natural light.

Indeed, one of the original wishes/intentions was to have a few skylights, but this was then dropped halfway through the project for various reasons. But the main point is that they want to control the lighting of the exhibits, so they required a museum that could be as dark as possible so that the lighting could then be artificial. So, the question has not arisen.

JAF – But there are issues of transparency; there's at least two large windows that offer a view of the river, a visual contact with the outside.

PMR – There are issues of illumination, but…

JAF – That's precisely my point: there isn't much light, but there's the basic light.

RF – There are two fundamental areas with natural light that mark the axes. There's cross-lighting, marking the axis that crosses the museum from one end to the other, with windows on both sides, though not aimed at natural illumination, but rather at marking this axis and, at the same time, providing a view over the river for those inside. It's a moment of pause in the pathway of the visit. And it's an orientation.

Then there are these two windows on the two ends marking the end of the museum, its extremes, and they're two narrow slits.

As I say, the aim was not to provide natural light; both were intended to mark moments along the journey through the space.

As regards the HVAC systems, it's obvious that spaces of this type have air conditioning demands, particularly because of height (inside, there are ceiling heights of 8 m, although the beams have 10, because you still have the truss thicknesses). And here, complex air conditioning and comfort issues arose.

Although we actually had space to do everything (because we have a 20-metre span, we need 2-metre trusses to carry anything we need through there), given the level of comfort intended, what we did was to choose two longitudinal lines, because the space for machinery – which had been more or less thought out beforehand as being placed in the roofs here in the middle – easily allowed a very clear layout to be defined along this wall. On the other side, it wouldn't be as easy, and so the solution we came up with was cooling piping and underfloor heating, and the combination of the two provides comfort in all seasons.

In any case, this integration is quite clear in situ, I think. These two huge lines that stretch the full length of the museum enable this distribution to be made.

PMR – The coaches, you see from here the bridge and the passing trains. And from here, you see the pedestrians walking past over there.

There's something interesting that I'm quite fond of (excuse me if I say 'interesting,' but I always like the theme of architecture, a good design, let's say). Suppose the original idea is a good design in its broadest sense. The intention starts to speak for itself.

If I lift this museum off of the ground, but have the workshops downstairs, however little I want to lift them off the ground, I have to lift them enough so that the coach can enter the workshop. And the minimum is roughly 4 m, because the workshops are here.

This means that the museum inside has a ceiling height of 10 m, and here, in its core, where the daily maintenance workshops are, it doesn't need more than the 4 m it already has below. This means that this whole pavilion with 10 m can be made with a ceiling height of 4 m, open on the top for all of the air-conditioning machines, distributing everything. And thus the design is all done.

BR – The integrating vision we talked about…

PMR – On paper! Then, you make better use, because you don't need so much there, so you have the second floors inside here.

That is, what you adopt as a design idea for each case has to be good, not to say optimal, from the outset, otherwise you have to change the design, what we call the architectural concept.

RBG – I'd like to say something on this issue of natural lighting and the relationship of the building with the envelope.

What I think is quite important is that, as the building is the response to a brief that goes beyond the museum's brief, that is, to respond to the whole issue of the envelope of the immediately surrounding territory, the openings in the building are also part of this reasoning. They're not there just to mark axis points in the building, but also to maintain awareness that they are the moment at which, from the inside to the outside, the construction choices in regard to that territory are recognised, meaning an engagement between the people inside and the people outside, and the whole city emerges in these two dimensions. That's why the openings appear as a strategic attraction. This attraction is not just so the visitor can see outside from the inside, but is also part of the understanding of the whole place and the salient features that gradually emerge.

PMR – I wanted to say- excuse me, because this is a technical issue, and even an ethical and political one in regard to the project, so I must mention it.

When setting up this group, I also have my support practice in Brazil, where I work with a group from my company, the MMBB group.

So, I retained Ricardo, the MMBB, Rui and myself. That was the team.

JAF – That's very important for the next point, the working methodology.

BR – Having established that 90% of the technological questions had been answered with this model, what information did you pass on to your team? I mean, and this will seem like we're teasing you again, how do you pass this information on to an external team that's not working with the others on a daily basis, dealing with the bothersome everyday problems of the project? Do you pass it on to Brazil, to Ricardo and to Rui?

Is it when the design and the ideas have to be turned into a document, into a concrete design (drawings, regulations)?

What I mean is: is it more comfortable for you to pass the boring part, so to speak, on to the outside? When does this moment happen?

PMR – That division doesn't exist, the boring part…

BR – I know, because I'm an architect, I know what the boring part is, the regulations, to have to print all the drawings, there's always a bit of time-consuming work…

PMR – There's no such thing. Nor is that the role of the team.

BR – Precisely, that's not what I mean! They're not there just to do the dirty work. How do they participate in this work if 90% is already completed? Is the 10% the documents?

RF – Wait, that's 90% of the design, not 90% of the work!

Of course, from the point of view of the work and the detailed development, knowing Paulo as I do, someone I have the greatest respect and fondness for, that's a heavy part of the work. What I think, though, is that the issue, in Paulo's case, is very pragmatic and practical. Paulo hasn't got an operating office, so he needs to work with external people. Not that he needs to, he just does. It's an option. He works with outside people who obviously put in hours of work.

PMR – They're not outside people any longer. I've worked this way all of my life!

BR – And they're with you every day?

RF – Yes, during these stages, they are.

PMR – I don't know whether this is interesting, but they're basically my employees. I'm not cut out to be a businessman. They're my associates at my invitation, if you like. So, it's a whole process of seduction and you don't know how it started. That's my view. Before they realise it, they're all committed to the project.

BR – It's not like a business, it's collaboration. Teamwork.

PMR – I don't know how to put it. Once again, music perhaps. It's the orchestra, you can't take anyone out. They're all in it together for the whole piece.

RF – And the process that develops from there is a really fun discussion process, very rich, from the perspective of experience, detail. You have to understand this: Paulo has a very clear idea in his mind of what he wants, but he agrees to discuss it, and we're all left there saying silly and nonsensical things.

When you look at those schemes, you see lots of drawings, and Ricardo draws perspectives, and Paulo doesn't like perspectives, he only likes sections, and he's mad at Ricardo because Ricardo only does perspectives, and so forth… Then, in the end, they come to solutions that are probably much richer because they include information from everyone and this process – which was for me, above all, great fun, fantastic in terms of experience – is the process Paulo likes to work with, because he clearly loves the discussion.

PMR – Whether we speak nonsense or not, I like to amuse the others!

BR – When you spoke of the experience of a jazz orchestra, comparing yourselves to jazz players, with the input each one gives, the difference is that there isn't exactly a conductor. The music is built on the technique each one puts in.

JAF – Jazz isn't just that. That would be a jam session!

PMR – It's as if it were…

RBG – This working methodology has a great virtue (besides sparing, in fact, some of the nuisance factor, but that's not the priority): when the team meets, it's to devote itself exclusively to that work. When you have a large studio with a team in place, everything is mixed up when you try to find the space and time to work on a specific matter. Often, the work is contaminated and merged with a number of other things that are in progress, happening at the same time.

The virtue of this methodology is that when this team meets, it meets for the project. There's a capacity for involvement in the project's work that's much richer because everyone who comes to work on it at that point is separated from other concerns they have and focuses exclusively on it. And this is a secret. It seems obvious, but it's very different from what we do in our projects with a studio of 50 people, because we're always being contaminated by other matters, by other things and other problems. This facility to find the space to devote ourselves to a job is essential.

This is why we speak of the pleasure of working, because we all undertake to build a time-territory exclusively for that work.

RF – You've simplified it a bit: it's more than that! It's that as well, of course…

RBG – This makes a huge difference.

RF – Clearly, it's also that, but actually there is a tremendous stimulus in the discussion promoted by Paulo. Paulo has this very interesting facet: he manages to have us all speak nonsense with serious results in the end, but during the process, everything is permitted. It's one of the things that excites me the most, obviously. And we've all said our fair share of nonsense.

JAF – I'd like to ask a question to do with things related to regulations and official bodies. Was it hard for you to work here in Portugal, bearing in mind the connections with entities like IGESPAR and others?

PMR – I've always been very well received and supported. I had no more difficulties than those that come with the job.

This place must be looked after with a lot of care and attention.

On the other hand, the demands are very clear. The train, the height, this, that, are demands arising from the city itself. We were very concerned with getting to know about them so we could meet them, and never say "Let's do something else!"

Some aspects, like the public car park in the front, by the river, represent a way of suggesting methods for solving the difficulties we encountered. The car park is important from this technical point of view. If there were individual parking places – for the cafeteria, the annex, the administration – the problem would only get worse, because it would require potentially unpleasant works, excavation, suppression, often repeating ingoing and outgoing ramps.

If you limit yourself to just one place, other qualities emerge: the port is served through the walkway, and the pedestrian landscape is enhanced.

And at the same time, because it's a large building, no one would pass up completing it on top with a horizontal floor, which would make up a 2000 m2 hall, where the most beautiful feasts in Portugal could be held, with the leisure boats sailing past in front, the best Christmas Eve party, the best wedding reception, you name it, with guaranteed parking, access by water, by train, by car. You turn the problem into a quality.

It's what we always say: turn what is apparently a problem/difficulty into a quality. Like the cutting of the cathedral stones.

JAF – But the idea is there; it may not be built immediately, but it's there and it brings it all together. This project does make the city.

RF – What happens is that here, for reasons no one has fathomed, the Lisbon City Council has rejected it. Right now, as long as things continue as they are, it won't be done. Maybe one day, but not now.

The idea of this car park was not only to cover the coaches. It was to cover the coaches and the whole area. Hence, this walkway…

JAF – And also to bring the river closer to Rua da Junqueira, where there is a disused car park.

BR – Still on the team's integrated vision: what have they made you change from the original model?

PMR – In this method of work, when the work is completed, you no longer know to whom, what, or when something happened! It's hard, I can't answer that!

I can't recall any conflict, or heavy discussion… I don't think there was one. The idea of a good concept is precisely that it can lead, because, like in jazz, the one who apparently breaks away from the project knows what he's doing with the others.

And if one does this, the other will smile, because he knows how he's going to respond. So, the image, contrived perhaps, a little playful, comical, of jazz, may be appropriate. You know who you're talking to. It's healthy, more or less like our conversation.

BR – Why did you let them talk you out of it, when you wanted to build those orthogonal feet, and they made you adopt non-orthogonal feet and deviate from the rule you had imposed?

JAF – As far as I know, it was a discussion! You're speaking about the shape of the administrative building's columns. What is in place there, from that working session, that jam session.

PMR – It's always been like that for me…

RF – There was a point when Paulo wanted to change the columns, but we were already so excited about them, we talked him into leaving them.

PMR – Ah, well. The first idea was consensual, because it was nicer, more elegant. Not least because those buttresses establish and ensure the very interesting transversal bracing, almost necessary from this perspective: if you put a vertical column there, it would have to have a large cross-bracing for that reason, and it would become a large column. If you solve it with those inclined cantilevered beams in the direction of the stresses/strains, with stresses, you give in. Nothing better. The structure loves to behave like that: to accompany the stress lines. All the more so because they're not the same. They haven't got a single common point. That is a very elegant way to show the type of stress being dealt with.

JAF – There's a connection between the structure, its function and its elegance. There's an adaptation to the stress that creates an image of elegance that is perceptible in the design.

Although it is orthogonal, perhaps it could even be seen as a certain form of brutalism of minimalist architecture, but it's not, because it has the beauty, the poetry of the structure with this elegance.

PMR – Quite true.

You want to see something that the whole world should know about, but is rarely commented on (because it's there to see)? When you're sailing in the middle of the Tagus, you see the bridge there and a train crossing the bridge. As this bridge extends a lot, after the suspension, into the land, it's possible to notice this: when you see the train entering the bridge – and the sailing speed allows the whole train to be seen as it crosses the bridge, from your vantage point in this scenery, you see this: the train passes at low speed and then, as it leaves the suspended structure, it picks up speed. In other words, if the engine driver wanted to topple the bridge, all he would have to do is cross it at high speed! Because of the load distribution that the train creates when it's moving over the bridge, it must travel at that speed to distribute the loads little by little, otherwise the bridge would collapse. So, what is apparently stable and final is absolutely dynamic!

And this is true, in general terms, for all structures, but you don't see it much. Hence, the beauty of a stadium, for example.

In general, these stress points, like the figure of the column we're talking about, the way in which the aerial structure of the museum rests on the large columns that are connected to the foundation, it becomes quite apparent that this is necessary. Once it's done, it makes greater demands more acceptable than mere simple solutions, like, say, a column every 10 m, or something like that. It's better! What seems difficult, on the contrary, becomes something better. To concentrate the loads, provide a wide dimension for horizontal stability, all that…

It's a lesson we can only discuss properly by looking at the figure. They're languages. Architecture is a discourse, first and foremost. Don't you think architecture is a discourse? It's a great discourse about knowledge. Born out of necessity. And urgency sometimes!

BTR – What is your role and the team's role in the various stages of the project, in the initial brief, the preliminary study, the final design? Then, an issue I'd like you to develop further: what is your position in the work? How was all this developed?

RF – The intensity of these jam sessions we've been talking about is stronger in the beginning, of course, so as to bring it all to fruition. And this intensity was also dictated by a specific reason: in the contract discussions, the design was approved, they really liked it, and they defined a budget. They said: "This is how much money we have to do it". About 31 million. They wanted to build a clause into the contract that would make it compulsory to respect the budget.

At that time, because there are contracts, responsibilities, etc., in addition to all of these discussions, we all agreed amongst ourselves that we couldn't sign the contract with that clause without knowing for sure the amounts involved in the work, the project costs. Based on this, an initial phase was defined, a month or so long, during which we would make this study – as a first installment of the design – we'd come to a conclusion and then the design would continue, or not, according to the values we arrived at. With regard to the structures, for instance, there was no previous experience that would allow us to know the costs of this solution; you can't use the costs of a different job to say that this will cost the same as that other. There was no previous experience!

This led to a very intense period of work, a month or so, with a large architectural team, with Ricardo, and our team, with all the trades, to produce a kind of preliminary study that confirmed all of these assumptions. This was a period in which lots of decisions had to be made and we had to make numerous contacts, so that we could define everything. Ricardo with the finishes, as he had to prepare an estimate for all this, and ourselves with engineering. This first stage was very, very intensive, and it allowed us to come up with not only a price, but with the solution for the whole design, basically. That's what happened.

JAF – That's great, because it's really a budget test. It should be compulsory for every project.

RF – But it was promoted by us, precisely because we didn't want to sign the contract without this knowledge. And it was mandatory because Paulo was about to take responsibility for something he couldn't control.

PMR – This method has one great virtue, which concerns my working method, which is the same as Rui's, as a great mathematician, someone who mainly calculates structures. He knew how to set up – and saw the need to set up – an office that solves the whole technological side of the construction. So, the interfaces between piping, water, electricity, compressed air, air conditioning networks, he coordinates all of that, and brings that desire to fruition.

In a sense, you can say that the project was always sustainable; these demands haven't resulted in much negotiation. It actually continued more or less as it began and, looking at that budget, the work, certain quantities decreased, rather than changing the concept.

RF – Yes, in the end, there was a need to make adjustments, particularly in the building area. I don't know if you've noticed, but Paulo has always talked about a length of 150 m. Up to this day, I've never heard Paulo acknowledge that the building only had 126 m.

PMR – A 20-metre pavilion ended up at 18 metres!

RF – Ah, he knows after all!

PMR – I know, but since it hurts a little…

JAF – There's still the issue of technical assistance that we'd also like you to comment on; how has support for the work functioned?

Let's talk freely about the design stage and technical assistance.

RBG – As for the design stages, what's worth mentioning is that we've always been together, as representatives of the three offices involved, and with MMBB in Brazil (in the latter case, more at the beginning). What is important to say is that because we had a solid basis, which is the matrix in this plan, almost all of the major issues in the building, even the technical ones, were solved. If you think about the air conditioning infrastructure, it's in this plan. Everything is here.

Where's the structure? It's here… Where are the room plans? They're here. Everything is already here. In this plan, it's all here. Everything was made quite easy.

PMR – I only came here as a tourist. They had everything under control!

RBG – Thanks to this first stage, what eventually happened was that we were obliged to have a preview of practically the entire project, even of the constructive and technical systems, from the word go.

Afterwards, there were always consecutive working procedures, on the technical and execution sides, more between Rui and us, in order to keep cross-referencing information: in the case of AFA, it was the investigation into the systems to be adopted, and in ours, on constructive systems, finishing materials and also detail.

We always had working meetings, even to discuss things such as the glazing, the types of span, the building finishes, etc. It always resulted from our talks. What was really funny was that the physical production centre gradually relocated from Brazil to Portugal, in a more or less natural way, that is…

PMR – Between architects, not in a very bureaucratic way…

RBG – Always between architects, in a natural way. It started as a series of working sessions in Brazil, then, during the transition, for the preliminary study and the licensing, it all moved to this side and, during execution, the major working meetings were held here in Portugal.

PMR – TAP helped a lot!

RBG – Somewhere down the line, inevitably, Bak Gordon and AFA appointed, in addition to our group, a project manager, who also coordinated the technical teams, and they worked in perfect concord, whether it was Engineer Armando Vale from AFA Consult, or Nuno Costa from Bak Gordon, and they spent all these years online, every day, cross-referencing architectural decisions with decisions from the different trades… Then, the exhaustive inventories of constructive systems, options about the large spans, the roofing systems, the skin zips, the interior finishes, the working systems for storage and workshop spaces, and so on.

So, a series of topics was investigated, but always in a very natural way. When a doubt arose that we thought had to be more specifically referred to Brazil, because it entailed some alteration – nothing specific comes to mind, but if it were the case, the contact was made.

What was good was that, from the start we all agreed more or less that this construction should also materialise through the constructive systems that we all wanted. Here, there was no problem, quite the contrary, because the intention with this building was also that it wouldn't rely on the choice of finishes or constructive systems, which need to speak for themselves. It relies on this. All of the choices of the final design should be optimal so as to match seamlessly with this intention.

BR – This integrating vision we were talking about at the beginning of this interview moves from your orchestra-like working methodology to the office working methodology. I mean, in addition to this permanent coordination between Bak Gordon and AFA, it also reflects on the way of designing – at the level of the BIMs, the Building Information Models. Bak Gordon didn't use them, but AFA did.

RF – We did. The BIM was used in the first stage of the project, not afterwards. In this initial phase of budget checking. The whole project, in this initial phase, internally at AFA, was created with BIM. It's these images – you only have the structure there, but these images exist with the mechanics, etc.

We only used it in this stage for various reasons, one of which was because Ricardo didn't use BIM, and thus the advantages its use had for the subsequent stages were obviously limited, so we eventually developed the rest from then on in 2D.

We carried out this stage with BIM – and we designed ducts, we designed the whole structure – and what happened was that all of the layouts that were solved in the initial stage were the layouts that were adopted all the way through to the end of the project, because any incompatibilities and issues were sorted out right then.

JAF – Have you used BIM essentially as a design tool? Was it helpful?

RF – It was a great help; all the problems were solved in the initial stage.

JAF – And from your point of view, is BIM an opportunity, a threat? How was the relationship with the contractor, does he use it?

RF – Mota-Engil has been using BIM for quite some time in their works, for various reasons. And not just Mota-Engil; Somague, for instance, used it in Casa da Música.

Engineers and contractors use BIM essentially for control and measurement reasons. That's the primary reason why BIM is used: always money. Because with it, they can better save and control. Really control the whole budget. Mota-Engil is quite experienced in it and wanted to use it from the start. We gave them our models, although they were already outdated, because the project had evolved in regard to certain details in the meantime.

For example, that one there is a 150-metre model, not 126 m. There was a lot to do to achieve it, and Mota-Engil used them for two things that served them rather well: they converted our 2D designs to 3D, using Revit. They started by noticing incompatibilities, which is another aspect of BIM. We subsequently found that they were detecting so many incompatibilities because their model was completely wrong. It was completely wrong, there were incompatibilities left, right and centre, because they were doing it incorrectly. But it was a good effort.

I'm an unconditional fan of Revit. I do think that, when the construction is all prepared, Revit is a fundamental tool!

JAF – Good publicity for a well-known supplier.

RF – But it's an important supplier! It could be Revit or something else!

RF – There's a piece of software that I got to know that is amazing, perhaps even more powerful than Revit. It's used by Frank Gehry, but it costs a fortune and the learning time is crazy. But it's very powerful, particularly for weird forms like those he does.

PMR – If it's a form, it can't be weird.

RF – In Singapore, for instance, designs are automatically checked. When you submit a project to the City Council for licensing, it goes directly into BIM and they have an automatic checking procedure for all of the basic rules. This actually results in an incredible efficiency across the building industry from the day it becomes institutionalised.

JPPM – In your opinion, to ensure some of this gain, who is the person who should require the use of the BIM? The architect? The coordinator?

RF – We're now involved in a process with Petrobras, and it's the Client himself who makes that demand; he says, "I want the design completely done in BIM, I will only accept projects done in BIM," but they're doing it seriously, they're not playing games. It's with a master model, done from A to Z, as it should be, everything completely controlled and regulated. In Brazil, that's already quite common, especially in these major companies.

PMR – These requirements and today's technology, which facilitates everything you mentioned, are really very interesting, because they once again pose the idea of architecture as a form of knowledge at the core of the issue, which is a political one: the making of the contemporary city, because this rigour lets you control everything, but you might find out that what needs to be changed are the laws!

JAF – That's the ideal cue for the final questions.

PMR – Since things may eventually start to go wrong, what you're building and the demands of the contemporary city, you begin to see that the issue is the laws. That's hard politics! The city, for us, exists before it is made. It's in the imagination. And this, in my view, places architecture once again at the heart of the problem, of knowledge, within the University. I'm thinking about teaching, from a young age. Everything needs to be reformed. Knowledge can't be turned into a problem. How do you do it? Exams: you pass, you fail! But knowledge has to be seen as a virtue.

JPPM – I'd like to know if it makes any sense – and we know the ideal would be the opposite – to adopt the BIM in isolation, even if the others don't want it. Was that the case here?

RF – In the case of engineering, it's very limiting if that's the case. In our case, taking care of all engineering projects, there can be an advantage. For people who only design the structures, notably concrete structures, the cost of making the model is relatively low. But if we're dealing with steel structures, it becomes much more complicated, because models become extremely heavy and hard to control. It's a very complex model to control.

PMR – Every screw is calculated, everything!

RF – Exactly. The design itself is more complex, more intense.

In my view – and internally we arrived at that conclusion and tried to act on it – we should progress to Revit, even if the architects don't choose this path. And at one point, economically speaking, let's say, we ended up by going back in terms of generalising the method and adopting it only for very specific cases, notably situations of a more complex geometry and with clear compatibility, because it's not cost-effective. Every change is complex, etc. It's not justified.

RBG – I haven't used it because our practice doesn't use this methodology and, until it proves to be an indispensable need, I see no reason to use it. I mean, the virtue of this model… I create a building section, create two building sections that tell the project's whole story. That's what I'm interested in. Revit, contrastingly, is very sophisticated, and then if they change the tying of two beams, the plasterboard level has to come down 3cm, and I'm in trouble again. I'm not going that way, I don't want to speak ill of the system, I'm just saying I don't work with it.

I'll tell you why it doesn't work in Portugal: in Germany, you prepare a design like this and it will be scrupulously developed from the point of view of execution, however long it takes. Then, it goes to the building site and will be scrupulously followed, and then, yes, it's a direct transposition tool to the site.

Whereas in Portugal, we all know that, with all constructive choices, beginning with the structural designs and ending with the door knob, the contractor always proposes the opposite and will always find a solution that changes your system. And you – however hard you fight for your system, when the work starts – and hopefully it will – you are already giving in 30%, and this 30% is work that's already been lost. That's it, I've said it.

The solution is[?] to make a BIM model for the execution and provide technical assistance full of drawings, which is what we do every day.

RF – What I think is that BIM has a difficult adaptation phase, but when this phase is overcome, it will become a widespread standard. And this experience with Petrobras I'm talking about is quite clear: the clients will demand it.

It's Petrobras managing, and since it's a building it will manage, it is interested in having everything well documented, the complete model. So, what will happen, in my view, is that it'll be the clients making demands in that direction when they find out about the benefits.

JAF – Especially institutional clients.

RF – That's what I'm saying, you don't work for Petrobras if you don't use BIM.

JAF – Since I'm an engineer, do you think architecture has significant weight, significant influence on politics, on the way societies are governed? Do you think architecture can still be a voice for the people today, in the sense of influencing, criticising, attacking, acting with regard to political leaders?

PMR – I think so, absolutely, but it's a slow, different process: a political process. The whole world today, from the perspective of architecture and urban planning, faces basically the same difficulty: the construction, the making of the contemporary city, the human habitat. Nature, the conservation of nature, which is talked about so much, is linked to the building of habitable conditions on the planet. Up to a certain point, despite it being an old image, the city is the ultimate novelty for us today. It's surprised us with some fundamental inconveniences.

Not helping the world, not foreseeing the continuation of our life, etc. Mainly in the face of problems that are not apparently connected to architecture in a very straightforward way, such as overpopulation, the relationship between male and female… The complexity of our life demands that we give what we call architecture another dimension, which is already there, though not clearly seen: the idea of architecture first and foremost as a political issue.

This broadens the role of the architect in society, in terms of his or her participation in decision-making instances, in ministries. The architect is not just a designer of projects, but someone who makes an impact on the political development of this requirement.

BR – But, in fact, it's engineers, not architects, who are in power, who have been important!

PMR – I know it's a rather silly way of saying something that's been said so many times before, but why not? It's the people, the voice of the people that has to be in power. And in particular, 'the people' considered globally, who clearly represent a new people. The difficulties are our virtues. These difficulties are our aims, they're there.

BR – This is a question I've wondered about since university, when I first became acquainted with your work: as a child, did you want to be an engineer or an architect?

PMR – I don't know. Because the question is so lovely, it would be a sin to answer incorrectly. Don't you think every child is a great engineer?

BR – Yes, of my three children, at least one is.

PMR – A child is an engineer: to walk, when they get up from the floor and walk to the chair to find their balance, they know they can only walk three or four steps.

We are aware of our condition, including talking; learning how to talk is a terrible effort. I'm increasingly convinced that education, school, especially for the little ones, is a terrible evil the way it is today.

We harm children more than we should. One needs a clear picture to be capable of self-criticism. The first thing a teacher says is: "Shut your mouth!" to prevent the child from talking. We must learn how to teach while children are talking. You can't tell a child to shut up. How can we create a school where the rule would be "speak up, say what you want to say" while we speak about mathematics, mechanics, etc.? The idea of elementary physics, for instance, the mechanics of fluids. Things that children know about. If they breastfeed, they know everything. Taking a bath, learning how to swim… It's done today.

We really don't know how to enjoy this terrible condition of ours, but meanwhile, we are helpless in the universe. It's up to every one of us to build ourselves.

I see this dimension, a concept of architecture as a form of knowledge. You cannot teach, only educate.

BR – Did you want to be an engineer or not?

PMR – Since Rui is here, I could say to him: I, through this business of architecture, found out the best way to be an engineer. The easiest one: how are we going to do this thing here? (laughter)

BR – But when you were 5 or 6, did you want to be or not?

PMR – I hardly know myself. I'd send any psychiatrist to the loony bin in just three sessions.

BR – Well, if you remember one day whether you wanted to or not, please let me know. Something you repeat often…

JAF – We all hardly know what we are. We really know ourselves as best as possible, we know what we are.

PMR – There was an advertising campaign recently in São Paulo on television. I don't really know what it's about, but it's a small child, a toddler, but it can already laugh. The ad goes like this: the father picks up a sheet of paper, playing with the child, and when the father tears the paper up, the child laughs.

Because this idea that a child destroys everything is the wrong way to build. The child loves to destroy.

JAF – We're really getting into poetry now, but that's right!

PMR – After all, a kaleidoscope is always destruction, everything is broken inside. It's a great pleasure to break established rules up.

The child knows everything, takes its shoes off… Do you remember a ride we took with Inês, she took her kids and they were all in pyjamas and they were not as little as all that. The transgression in the face of an appeal, desire, pleasure. It's fantastic, it would be the erotic view of life. That's fundamental. To feed the erotic dimension of life.

BR – My question about your childhood has to do with what you say in interviews I've read, that your father used to say from early on that technique had the power to transform.

PMR – My father never said that to me, but he would take me, whenever possible, knowing what he was doing, on outings out to sea, with dredgers, aware that I'd love to see the seagulls. In a sense, I was also looking at the machines. Yes, I learned that lesson early on. He loved it and was aware of the virtues of technique and engineering.

In São Paulo, my father has always worked for the government, for the Secretariat for Transportation and Public Works.

At one point, they wanted to modernise cabotage in the small ports, mainly to support fishing. Sometimes fishermen are lucky and haul good catches, but they have nowhere to unload, collect ice, and they must carry on fishing because the shoals of fish are going well.

What did the government do? They took a small boat from the existing coastal navigation company, the Itaipava. All the boats were called Ita-something: Itapé, Itaiti, Itaipava.

They took the Itaipava, they put everything the engineers wanted inside and they undertook a survey of all of these ports. They took 30 days to complete a job they had also done in a different way.

One or two engineers took their sons, and my father took me as well. What I saw for 15 days, docking at each port, was, amongst other things, in one of the ports, a small obelisk and an old cannon, one of those with a tube, in the main square. They decided that, when the boat came into port, they would fire a salute with this cannon. So, they rehearsed and fired, they were delighted, but they thought that it should be more powerful. The day the boat arrived, they put in too much gunpowder and a bit of the tube fell on the deck and exploded. Everything stopped. A rope had already been thrown, but the boat lifted it and reared up backwards. So, in regard to engineering things, I saw a big disaster due to an excess of eroticism.

This was in Cananeia, the small town of Cananeia, where the cannon exploded; luckily no-one died, because it fell outside.

The town was right there…

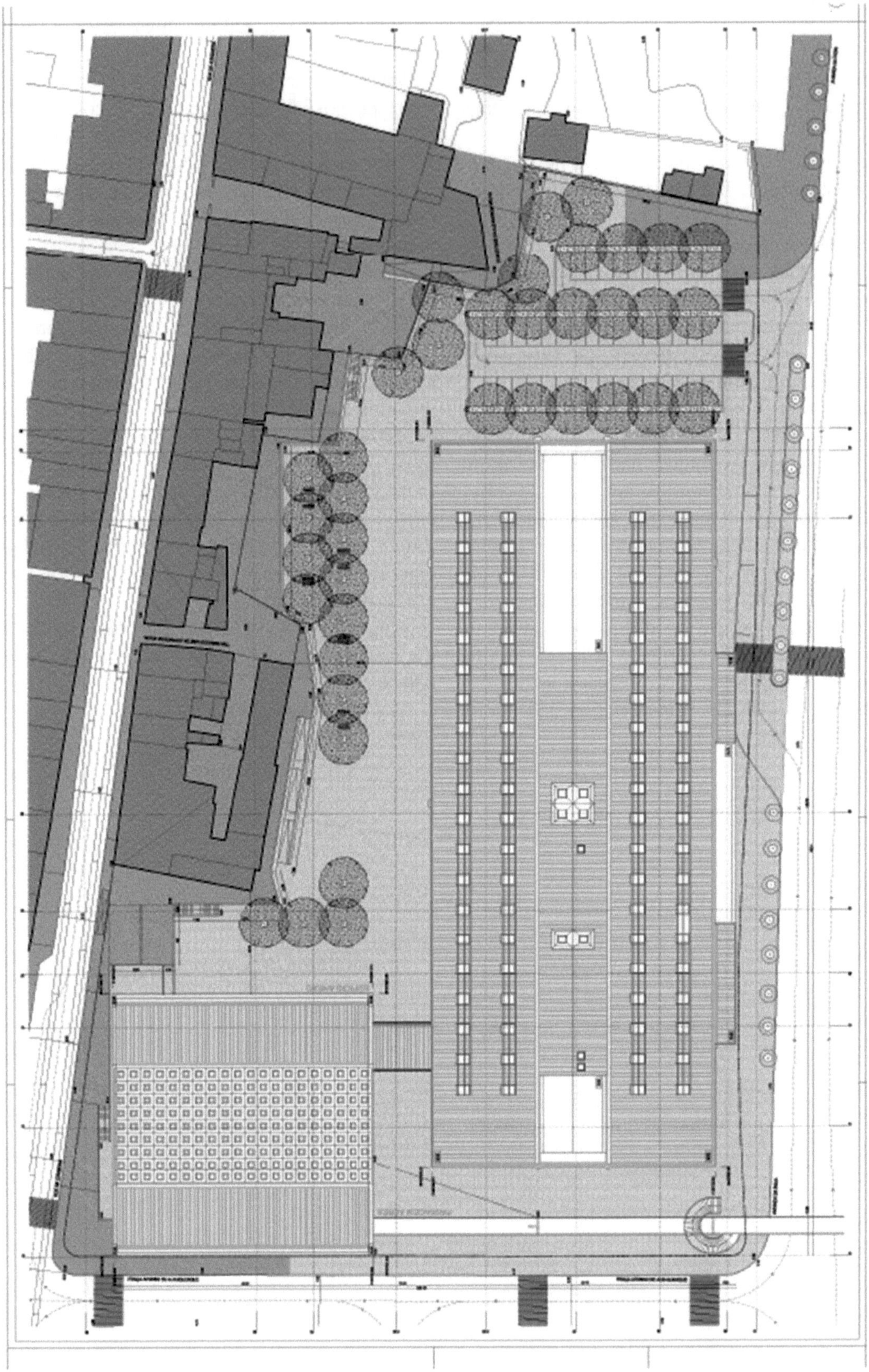

Coches Museum site plan

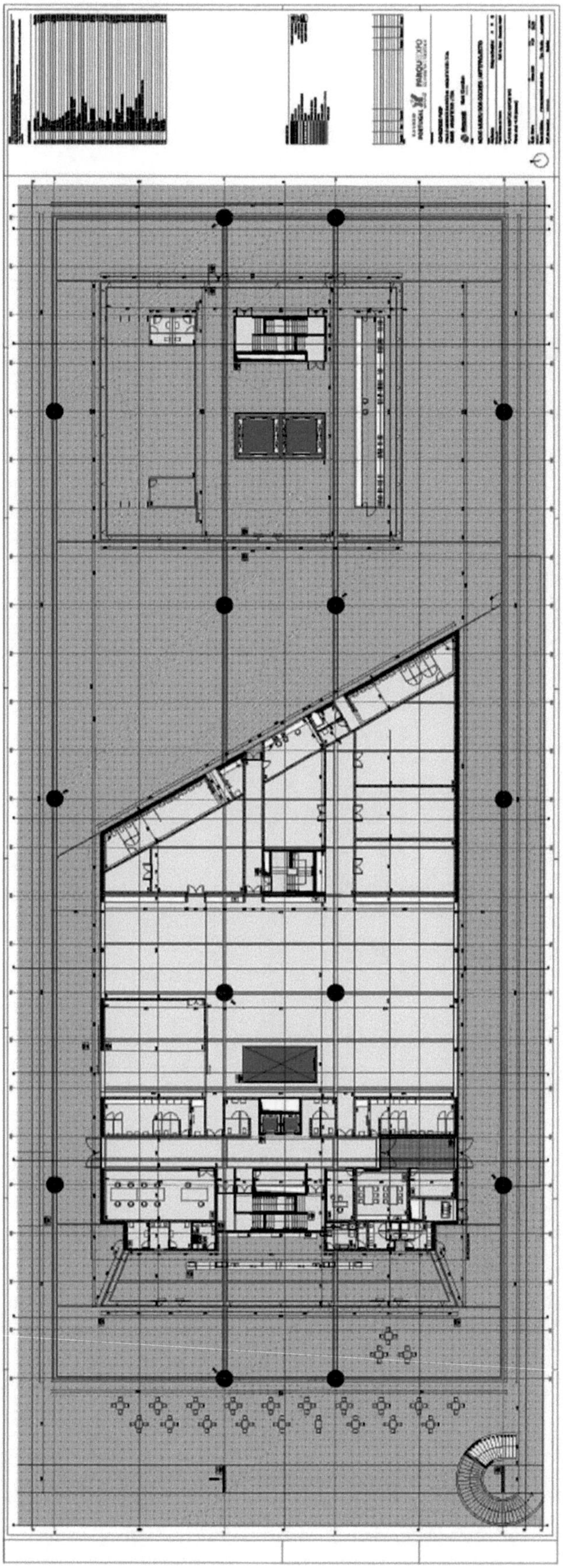

Coches Museum floor level plan

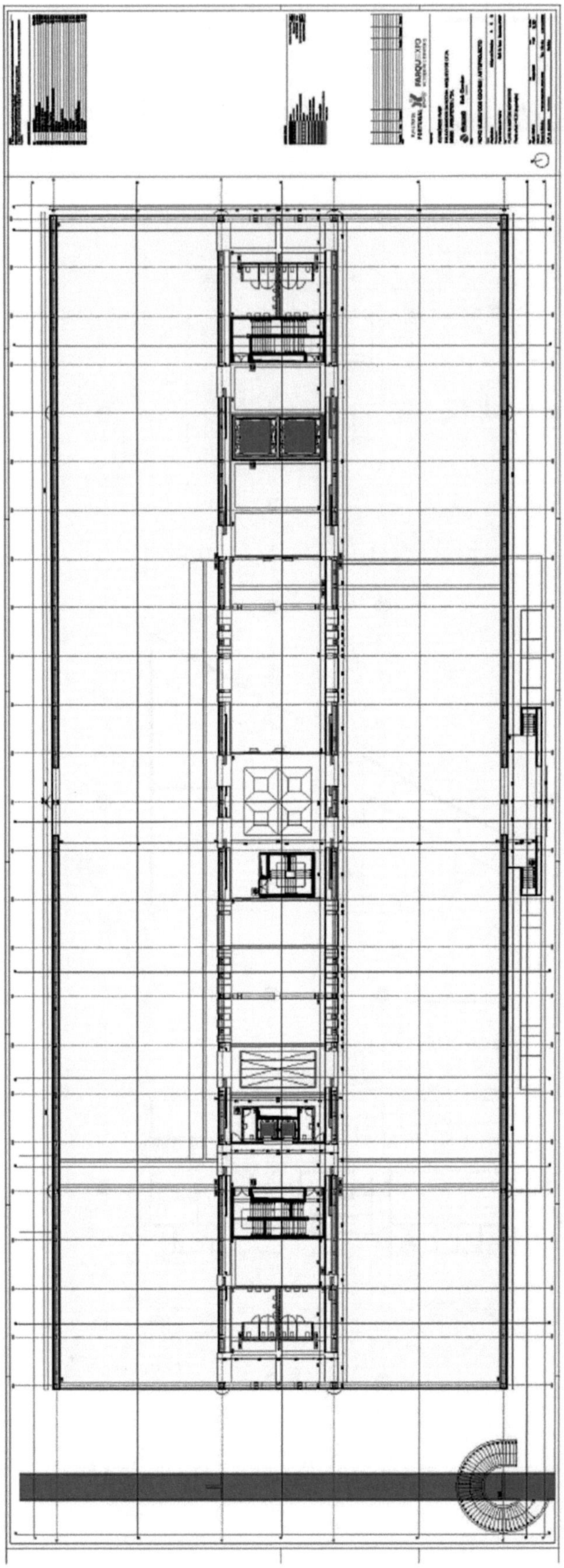

Coches Museum level one plan

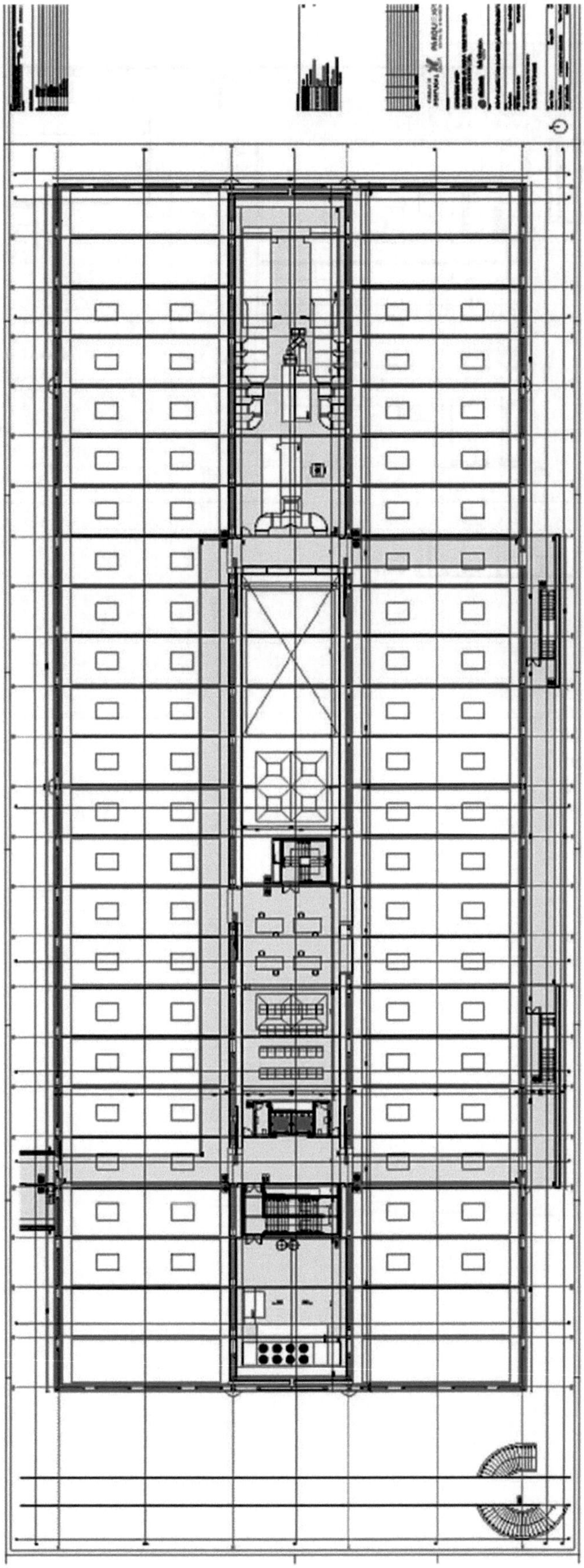

Coches Museum level two plan

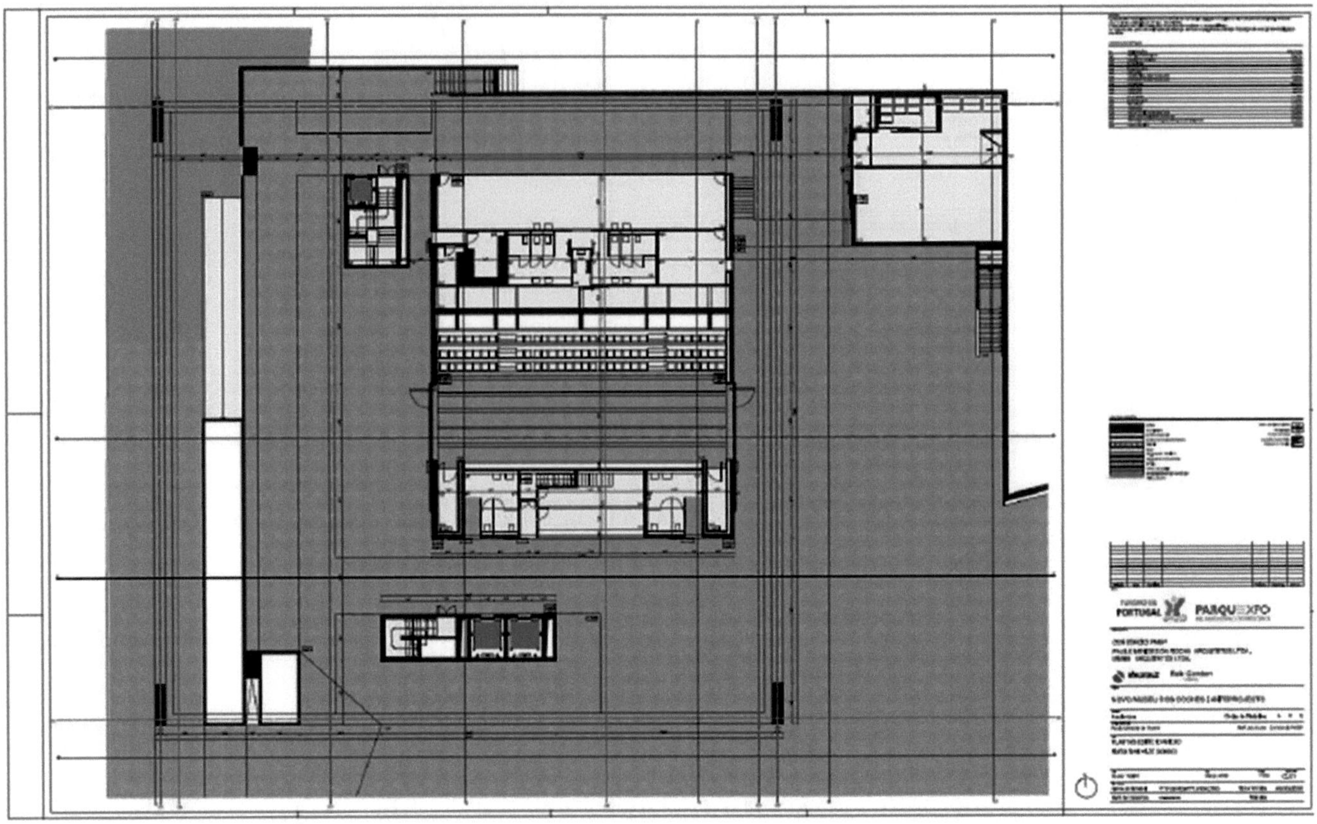


Coches Museum auditorium level one plan

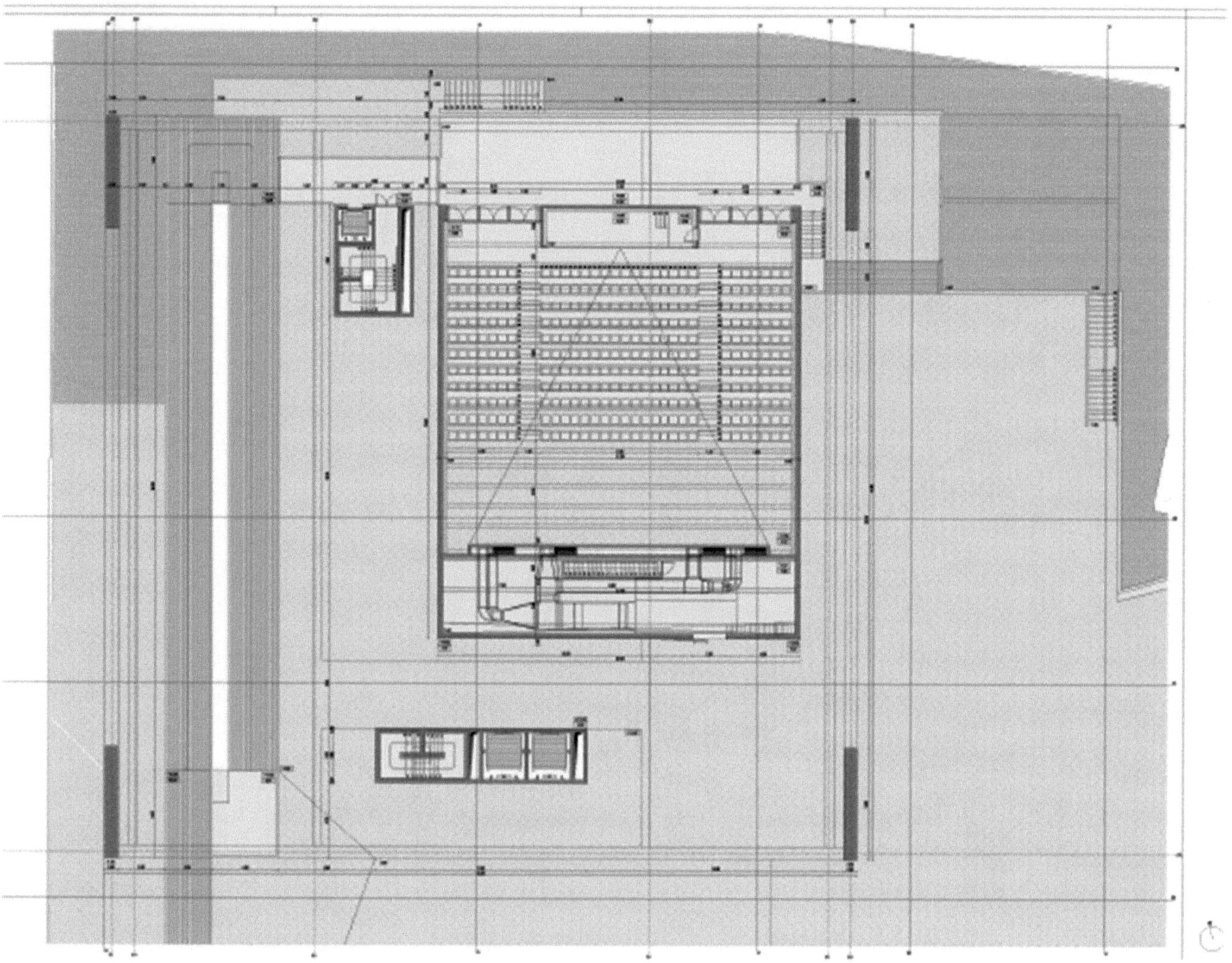

Coches Museum auditorium floor level plan

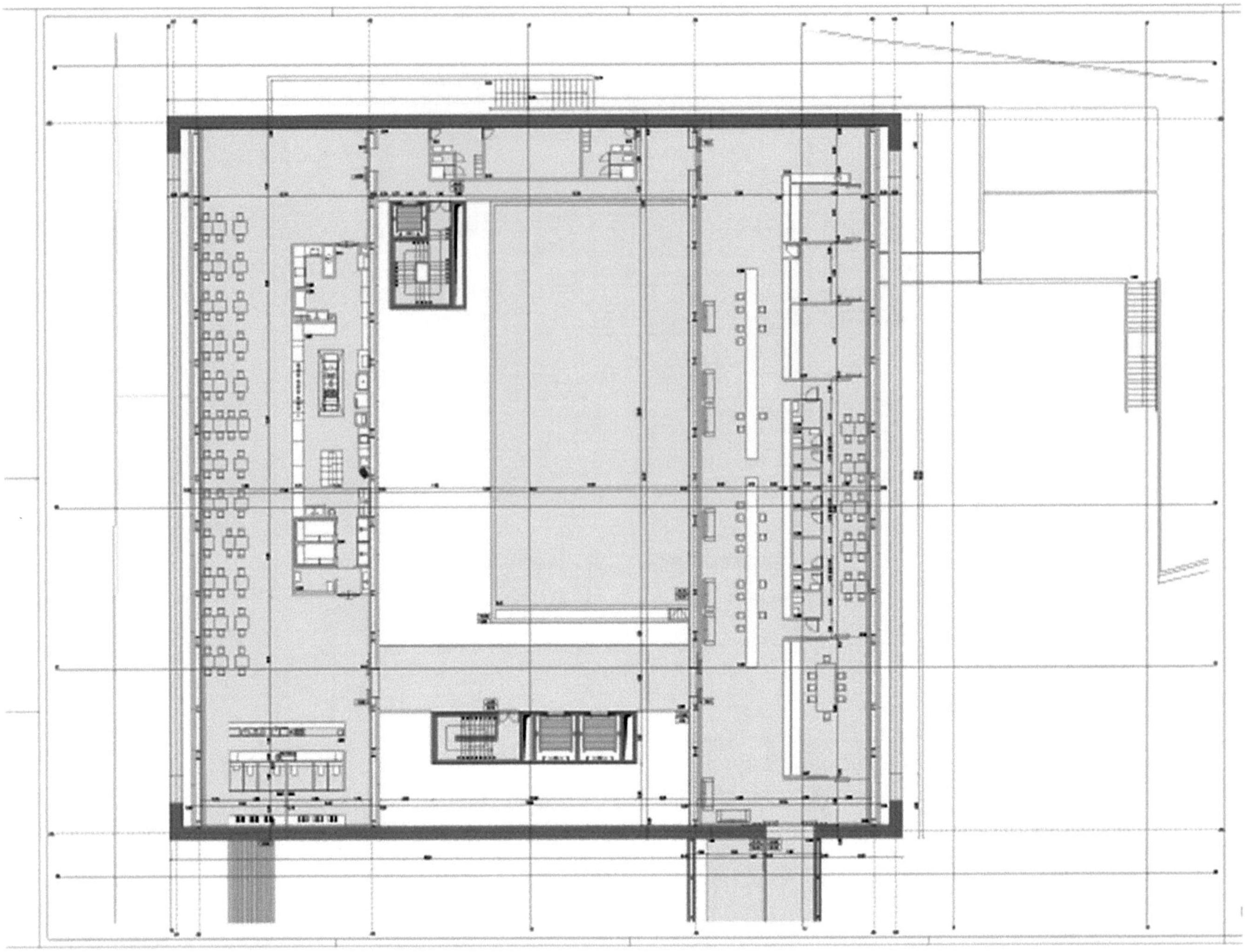

Coches Museum auditorium level one plan

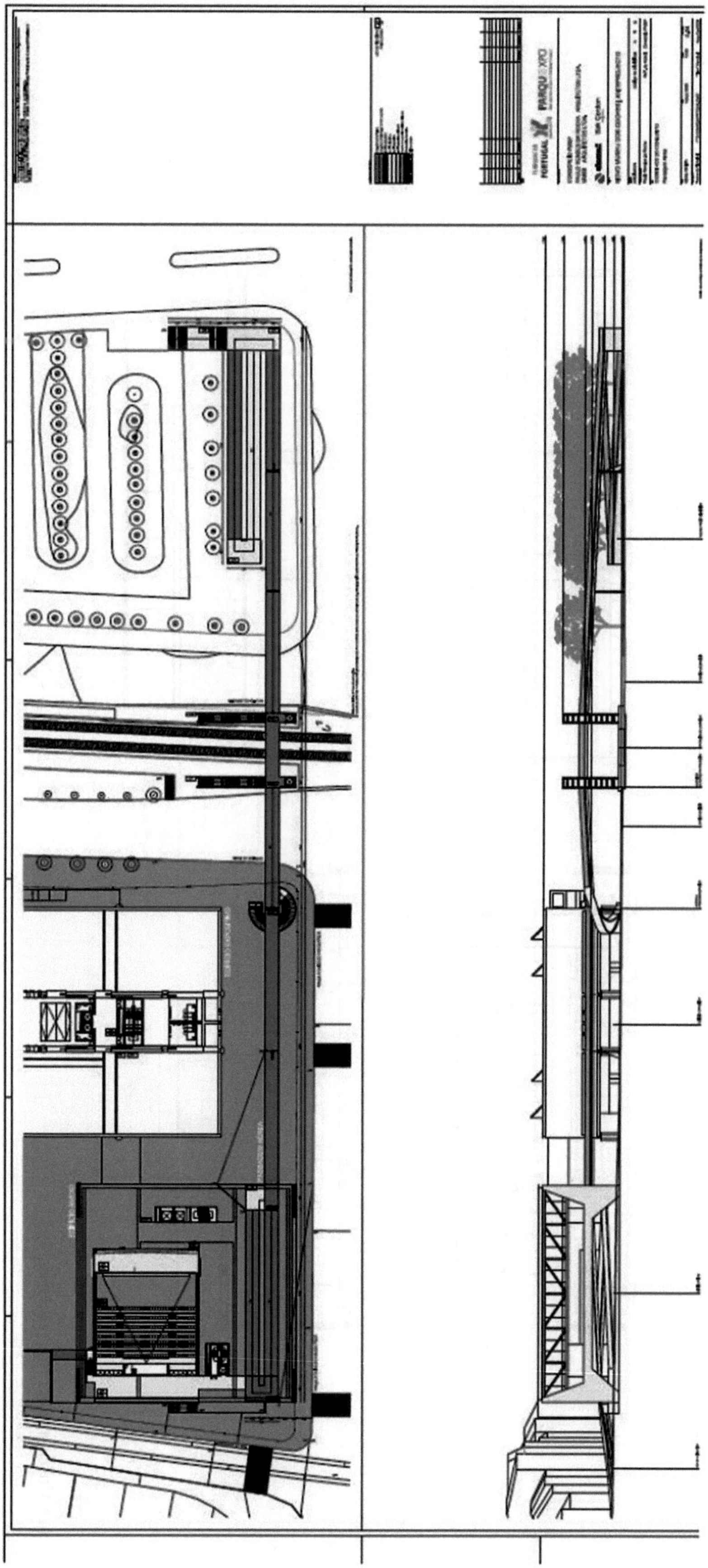

Coches Museum auditorium River front façade

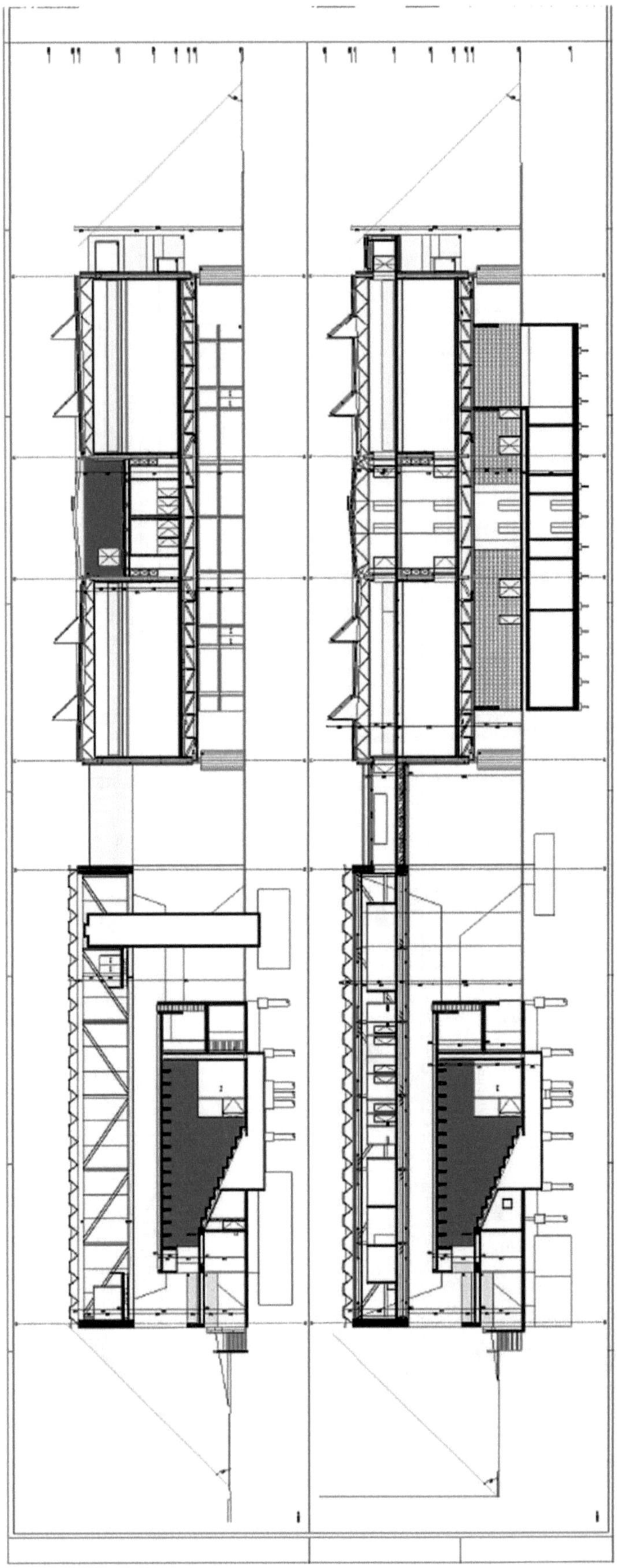

Coches Museum auditorium cross sections

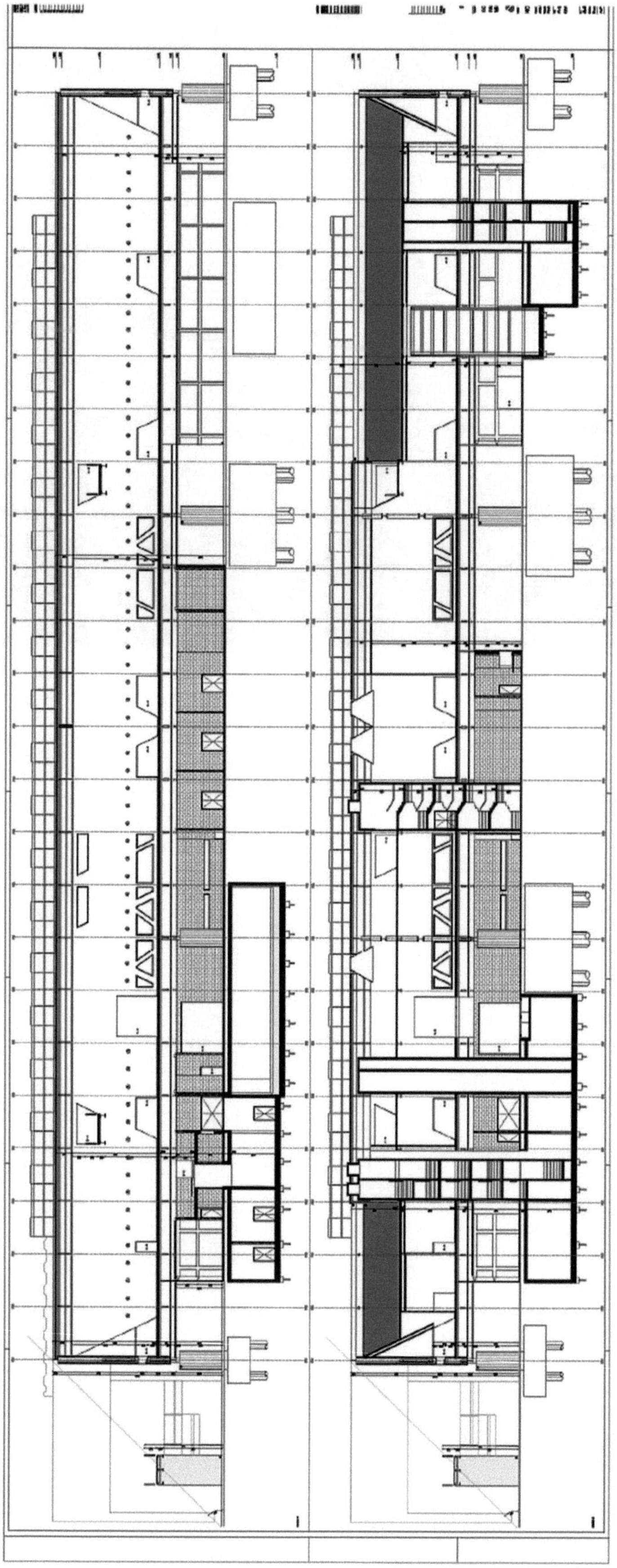

Coches Museum Long sections

Coach Museum
Externel view from Av. da Índia
Copyright Arménio Teixiera

Coach Museum
Externel view from the interior of the courtyard
Copyright Arménio Teixiera

Coach Museum Auditorium
Externel view from Junqueira street
Copyright Arménio Teixiera

Coach Museum Auditorium
View from the upper level facing the auditorium roof
Copyright Arménio Teixiera

Coach Museum
Internal view of the museum
Copyright Arménio Teixiera

Coach Museum
Internal view of the upper level facing the Afonso de Albuquerqie Garden
Copyright Arménio Teixiera

# The Engineering of the New Coach Museum

Rui Furtado, Armando Vale, Paulo Silva, Bruno Henriques, and Luis Oliveira

The project began in late 2008, following an invitation from the Portuguese Government to architect Paulo Mendes da Rocha. The goal was to open the Museum on October 5, 2010, as part of the centenary celebrations of the Republic. The New Coach Museum was to ensure the preservation of Portugal's Coach Collection, and display it to an audience of about 1,000,000 visitors each year.

Paulo came to Portugal to see the site and obtain information about the Museum's programme, after which he returned to São Paulo. Two or three months later, he returned to Lisbon to present his ideas – his concept for the Museum was based on the idea of a white "box" in which to keep the mostly Baroque "Treasure," profuse in ornamentation.

For the presentation of the Architectural Project, he basically brought photographs of his working model, and a clear vision of what he intended to do:

1. The Museum area – raised off of the ground, with workshops and a cafeteria; the entrance to the Museum would be located on the ground floor;
2. An Annex, connected to the Museum by a bridge, housing the Museum's administrative area, with a restaurant above and an auditorium below;
3. A pedestrian walkway linking Rua da Junqueira to the Belém River Station;
4. A car park – a continuous ramped silo – next to the Tejo River, completing the pedestrian walkway;

**Fig. 1** Example of old car exhibited in the museum

The peculiar spatial arrangement of the buildings, with the aerial link between them, creates a portico that marks the entrance to the internal square, where the buildings on Rua da Junqueira form a new frontage, promoting their revitalisation in a reconstitution of what was once Rua do Cais da Alfandega. This arrangement highlights the relationship among areas of different heights, enhancing the movement of passers-by, consolidating the restitution of the whole locale to the city, and strongly inviting a visit to the museum itself. This basically reflects what the Museum is, according to Paulo Mendes da Rocha's vision, a public place – "carefully protected and unexpectedly open."

The Proposal was a response to the brief, but extended its scope, in a broad reading of the site and its potential requirements, favouring the creation of a qualified public space.

Except for the details, it is surprising how closely the initial proposal corresponds to the end result – only the pedestrian walkway (its construction will start soon) and the car park silo (rejected by the Lisbon City Council) are missing. All the rest is there, just as Paulo envisioned from the start. A little smaller – for budgetary reasons – but the same spatial arrangement, with siting as in the initial proposal. Throughout the process, there was no display of prima donna stubbornness, nor was there a need for any engineering speciality to be consulted – all the building's needs were covered by the simplicity of the original idea – basically, everything had been thought through!

In the Twenty-first century, any building, but particularly one of the nature of the New Coach Museum, whose purpose is to protect a unique Treasure and to provide the facilities to

R. Furtado (✉) · A. Vale · P. Silva · B. Henriques · L. Oliveira
AFA consult, Vila Nova de Gaia, Portugal
e-mail: Rui.Furtado@afaconsult.com



B. Rangel et al. (eds.), *Museum Technology and Architecture*, Building Research: Design, Construction and Technologies,
https://doi.org/10.1007/978-3-319-76172-5_7

**Fig. 2** Aerial view

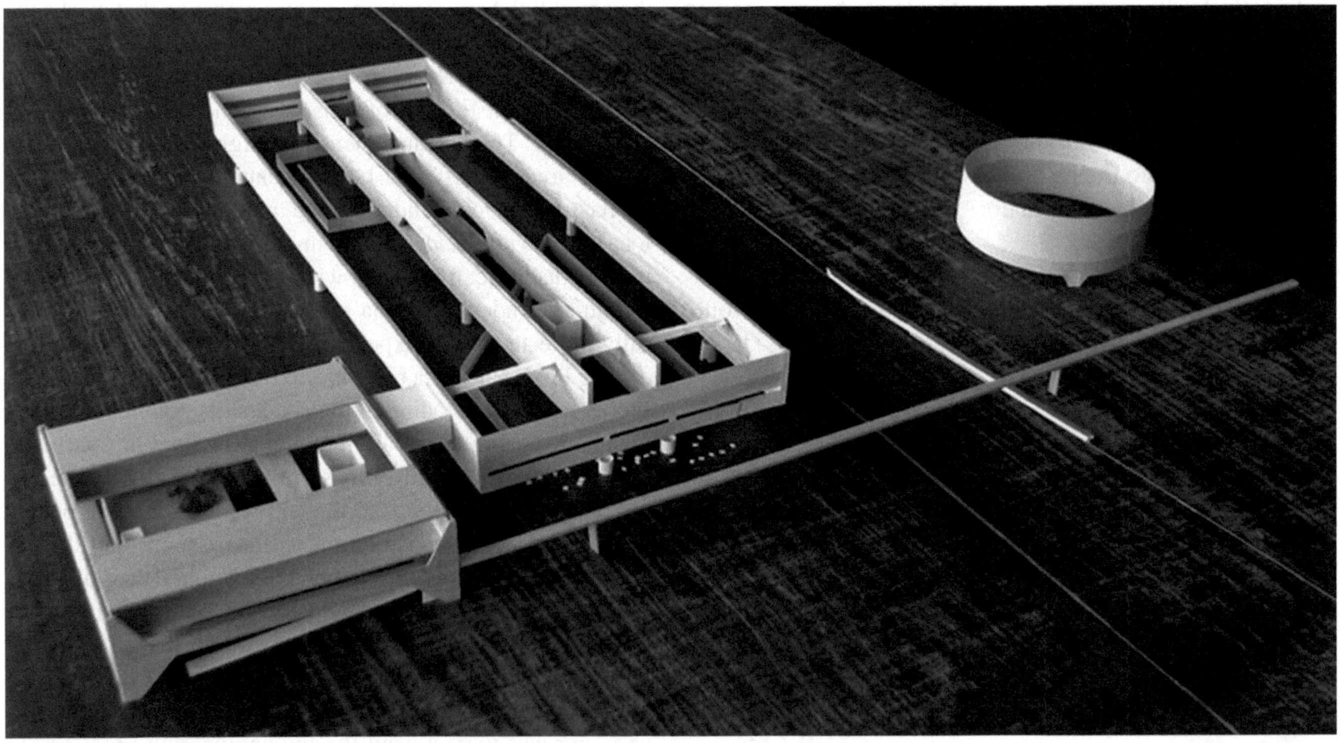

**Fig. 3** Conceptual model made during the project by Paulo Mendes da Rocha

receive one million visitors annually, is a machine whose efficiency and success depend as much on the way in which its various components fulfil their function (without prejudice to others) as on its ability to contain spaces capable of surprising and inspiring. This need requires that the Architectural and Construction solutions be adopted simultaneously and meet the needs and demands of the various systems necessary for the proper functioning of the building in equal measure.

Coming from Brazil – the land of "exposed concrete" – the building was envisaged as being made of concrete. However, considering the 50 m-long spans, the very short construction deadline as had then been defined, the plot on which the building would be sited, the need to take seismic activity into account and the requirement to accommodate complex and demanding environmental control systems, a switch to a lightweight system of construction was suggested – a monolithic metallic structure, "anchored" in the centre and sliding in the peripheral supports, and with lightweight plasterboard walls.

The proposed system of construction then arose as a complement to the architectural solution, allowing the mass of the building to be reduced and the cost of the foundations and vertical elements to be rationalised, while at the same time, guaranteeing a minimal construction period. Meanwhile, due to various setbacks – availability of the site, the transfer of services that were installed there, etc. – the deadline was no longer seen as a priority, though the other assumptions for the decision remained valid.

For the structure of the box, with longitudinal spans of 46 m and transversal spans of 18 m and 12 m, the option was for a hyperstatic, redundant truss structure, with three-dimensional behaviour, taking advantage of the main structural elements, such as the bracing elements in the main structures in the other direction, and making the most of all the continuities and stress redistribution that it was possible to create, always looking for a high degree of simplicity of assembly.

**Fig. 4** Main structure of the museum during the construction from the outside

**Fig. 5** Main structure of the museum during the construction from the inside

The large longitudinal walls of the box house the main trusses, which take advantage of the full height of the building. It is along these walls that the main arteries for the distribution of services are located, linking the Museum's two technical areas, located in an open space at both ends of the upper level.

The Architect imagined the Annex as a covered building, half empty, in which the concepts of inside and outside are blurred by the variety of views and environments that it creates. A portico structure of pre-stressed concrete supports two steel and glass boxes, where the Management offices and a Restaurant are located. The Auditorium occupies a concrete box at the ground level.

**Fig. 6** Final model

**Fig. 7** View from level 2 of the auditorium

**Fig. 8** Ramp under construction

**Fig. 9** View of the ramp under construction from Afonso Albuquerque Garden

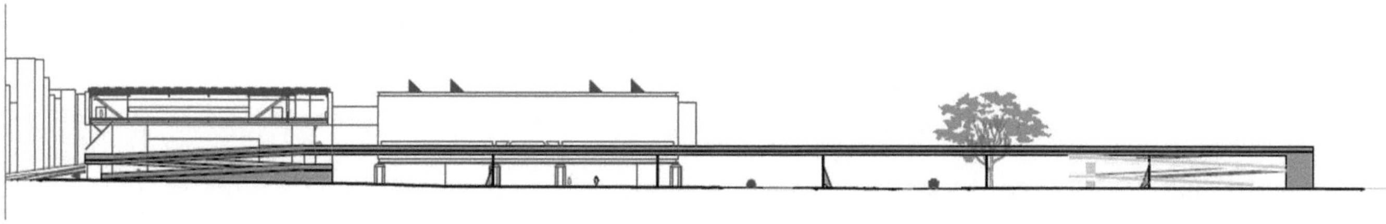

**Fig. 10** West façade

The ramp that links Rua da Junqueira to the aerial walkway that carries pedestrians to and from the Belém River Station ends in this Annex. Its ground elevation, originally constant at 6.60 m, coincided with the height of the Museum floor: this was a straight line almost 300 m long that connected the monumental area with the riverfront and the projected car park.

This walkway and the intended car park, whose capacity will be 400 cars, form a proposal for minimising the problem of parking in the Bélem monumental area, which is currently full of parked cars on the surface. This component of the project, which would be financed by the operating concessionaire, has been turned down by the Lisbon City Council.

The aerial walkway, which now respects the demands of REFER (particularly a headroom of 7.50 m above the railway line – about 8.1 m from the ground), was intended to be pragmatic, of simple construction, with a language that was "linked" to the Museum. 3 sections of 3 spans, about 30 m each, were defined, the decks comprising 3 parallel HE550A beams. The pillars were slender walls of reinforced concrete. The guard rails are identical to those in the Museum – solid pre-fabricated panels of white concrete, supported by metal uprights.

It should be noted that the construction of the Museum respected the budget initially set out by the Client and accepted by the Design Team, to the cent. The work has been completed and is ready to be inaugurated. At the time of this writing, tenders have yet to be launched for the Exhibition Project itself – essential for the opening of the Museum, as well as the construction of the aerial pedestrian walkway.

## Exhibition Pavilion

The exhibition pavilion is destined to house the coaches, which will be transferred from the actual Museum, and is located parallel to the Avenida da Índia. The building is approximately 16.5 m high, and could simplistically be described as a white, opaque parallelepiped, 126 m x 48 m x

12 m, supported on a grid of 14 circular pillars 1.80 m in diameter. Internally, this high volume is divided into 3 longitudinal aisles.

In addition to the raised floor, which forms the core of the museum, there are also certain complementary areas on the ground floor and a small underground area.

The exhibition areas are located in the lateral aisle of the upper level, and include two large halls for the permanent exhibition, of about 125 m x 17.25 m of usable area each, with a double ceiling height (8.28 m), marked by a continuous pavement of smooth concrete, pure white walls and the white metal grid of the suspended ceiling, which encloses all the infrastructure of the building, but, at the same time, allows it to be seen. There is a sequence of openings in the museum's longitudinal walls, whose shape relates to the configuration of the metal truss structure and, in the case of the intermediate walls, windows that result from the subtraction of their volume to the volume of the walls in which they sit.

Between these two halls, in the central aisle, there is a group of spaces, including a temporary exhibition hall, with about 215 $m^2$ of usable area and direct access to the entrance, some workshops for everyday maintenance, where the coach lift and toilets are located, and the public circulation space that links the various spaces and enables access to the emergency stairs, also installed in the central nave.

Level 2, with 1380 $m^2$ of gross area, is limited to the central nave and comprises, at its two ends, space for the technical areas where the ventilation and air conditioning equipment for the exhibition halls are located. There is also an education department room and a group of hallways and corridors that provide a route on the upper level, giving a more exclusive view of the exhibition halls, as well as access to the external balcony located on the south elevation with views to the river.

Access from the upper footpath to the Annex building can also be achieved in these corridors, giving Museum staff direct access to the administrative areas.

**Fig. 11** Roof structure under construction

**Fig. 12** View of the main room of the museum in the structural construction phase

**Fig. 13** View of the slab of the main room of the museum in the structural construction phase

The need for tight control of the lighting in the spaces allocated for exhibition means that there are practically no openings to the outside. Only horizontal slots are provided at each end of the building, with one or two other apertures carefully arranged along its longitudinal elevations. Skylights, used only for natural fume extraction, are provided in the roof of the exhibition halls.

At the ground floor level, areas are divided by two independent modules, one being the entrance and the other being for the reserved areas and the service area. The entrance module, which includes the shop and the storage lockers, is enclosed by completely glazed façades, creating continuity with the exterior square. There are two large elevators here, each with a capacity for 65 people, which provide access to the exhibition halls on the upper floor. The ticket offices and public toilets are in the other module, facing the entrance to the museum. The maintenance and restoration workshops are located in the centre, accessed from the outside through a large gateway that leads to an antechamber (reception) and a cleaning room. This then leads to a large open space where the maintenance workshops and vehicle storage areas are to be found, along with the lifting platform on which the coaches and other items are raised to the exhibition hall level.

**Fig. 14** Wall structure under construction

**Fig. 15** View from the exterior in the structural construction phase

**Fig. 16** External view from the east side of Av. da Índia

Certain areas are specifically set aside to store more sensitive items, such as saddles and cavalry accessories, textiles, documents, paintings and other objects. These are closed and have no external lighting. Other service areas are to be found adjacent to the workshops and on the side opposite the ticket offices, including showers and dressing rooms for museum personnel, staff rest areas, curators' rooms, offices and service elevators for access to the basement and upper floors. In the extreme west of this area, there is also space for lease, intended for a cafeteria with a terrace, and, as in the entrance module, it also has wholly glazed external façades.

The basement level is intended exclusively for technical and support spaces. Amongst other things, it includes water cisterns for fire-fighting, drinking water and reuse of rainwater, plant rooms for HVAC equipment, a generator and an area for the recycling and storage of waste.

In general, the floors are made of concrete slabs finished with surface hardener and mechanical smoothing. The external cladding of the façade of the pavilion is of the "ventilated curtain" type using Knauf "Aquapanel," the composition of which includes cement boards, waterproof plasterboard, thermal insulation, rockwool and a vapour barrier. The interior finishes of the façades are made of plasterboard panels and thermo-acoustic insulation. The roof is coated with a sandwich system comprising a structural plate at the bottom, a layer of thermal and acoustic insulation and an external profile of the Kalzip type.

Owing to the need to have large spans, associated with the concept of a building raised from the ground, we opted for the use of light construction systems, evident not only in the main structure of the building, which is metallic, but also in its entire envelope. Note, in particular, the composite slabs of reduced thickness of the exhibition floor, the sheet metal roof and also the interior and exterior walls, which are predominantly of plasterboard.

## Annex Building

This is a building that alternates between full and empty, which includes the independent module of the auditorium at its base and, above, the volumes suspended in the structural exposed concrete porticos for the administrative areas and the restaurant, with glazed façades adjacent to the triangulated structural beams.

The interior of the building includes an arrangement of vertical and horizontal communications, predominantly public, all in an outdoor environment, though sheltered by the central skylight constructed of glass and metal.

On the ground floor is the auditorium, 8.50 m high and with a capacity for 330 persons, built in exposed structural concrete to be painted on the external side. This is an informal space for visitors to the museum that also supports the educational department, with two large side gateways to allow passage of a horse-drawn carriage. Inside, concrete also predominates, the walls being clad with perforated concrete panels, behind which there is insulation that will simultaneously provide acoustic control of the space. There are prefabricated concrete T-section beams in the ceiling, and the floor has granite cubes at the base and precast concrete seats. There is a reflecting pool on the roof. Still on the ground floor, to the north, as part of the auditorium module, there is an area for a leased shop, and to the south, public toilets and a space for the Museum's reception and information desk.

On Level 1, at the rear of the auditorium, is a large balcony overlooking Rua da Junqueira, which also serves as a link between the new access ramps to the pedestrian walkway and Rua do Cais da Alfandega Velha.

Level 2 comprises two "bridges" in metal, with triangulated beams of a 45 m span, supported on the exposed concrete structural porticos. The eastern "bridge," with a splendid view of the new Museum square and the reflecting pool on the rooftop, is intended for the Museum administration, with spaces for administrative services, curators and management offices, library, toilets, staff kitchen and rooms for rest. The western "bridge" has an open space with a view over the whole monumental area of Belém, which is intended to be leased for a restaurant. This has a kitchen, bar and toilet areas, and the remaining space is for the dining area.

The connection between the "bridges" is made of two walkways with a metal structure and concrete decks, which also provide access to the exposed reinforced concrete blocks that house the emergency stairs and the elevator shafts.

There is also a small plant room on the north walkway.

The pavements of these spaces are finished in smooth concrete, the façades are glazed and the ceilings use a suspended metal trellis, thus maintaining the continuity of the solutions adopted for the exhibition hall.

**Fig. 17** Skylights of the auditorium under construction

**Fig. 18** Auditorium under structural construction phase

**Fig. 19** View of the 2nd level auditorium under structural construction phase

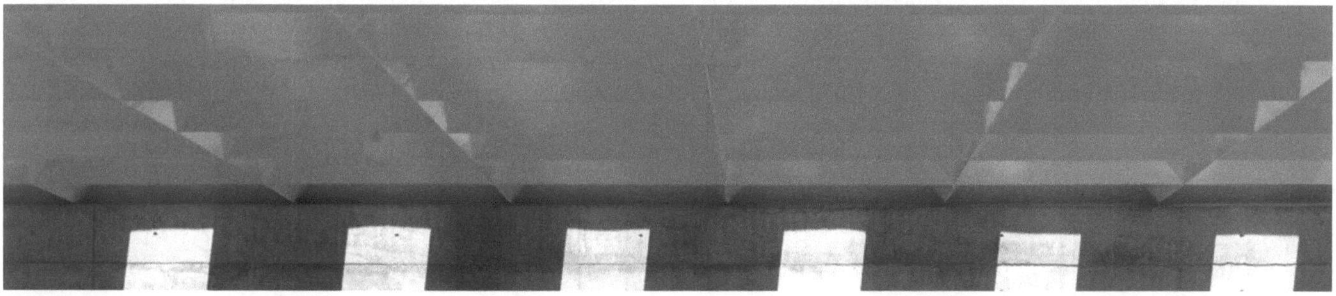

**Fig. 20** Skylights of the auditorium under construction

## Pedestrian and Cycle Walkway

The proposed solution for the pedestrian walkway retains the concept of public space that guides the entire project. Starting from the north, it is still within the confines of the annex building, attached by a set of 3 m wide ramps that lead from the level of the square to the level of the elevated deck, also providing direct access to the north veranda of the auditorium from one of the intermediate landings. After crossing Avenida da India, the railway tracks and Avenida Brasília, the deck elevations descend until they meet the ramps, again allowing access from the garden.

The height of the elevated part of the walkway is determined by the need to ensure a minimum headroom of 7.5 m for the crossing over the railway, as well as the need to comply with the maximum inclinations in the regulations.

In addition to the ramps provided at both ends, interim access by elevators and stairs is also provided, enabling the walkway to be accessed from the pavements on Avenida da Índia and Avenida Brasília, as well as access to the railway platforms in both directions.

## Square

The square under the buildings is made of a unique surface finished in granite cubes, occasionally interrupted by areas and elements of the building at the ground floor level. Its insertion within the envelope involves the resolution of the transitions between the different heights, which, in the area of greater slope to the North, is achieved by a set of ramps and steps and, in other areas where the difference is less marked, by the occasional gradual bending of the pavement levels.

The solution proposed focuses on the complete opening of the square as a public space, taking advantage of the permeability of the buildings at the level of the square and the absence of any other barriers to the movement of pedestrians. Vehicular access, although possible and necessary for both maintenance and loading and unloading, is closed to the general public.

Buses carrying visitors are able to stop and park to the south, on a road adjacent to the museum itself, and this allows visitors to be transported close to the entrance.

## Structure

The beginning of the project revealed that the structure would immediately assume a role of particular importance in the design of the building, indeed along the lines of the work of the architect, Paulo Mendes da Rocha.

The development of the project confirmed and extended this idea, and it was observed that the role of the structure was relevant not only in terms of form, but also in its contribution to the definition of finishes and architectural details, with the choice both inside and out of finishes in either exposed reinforced concrete or steel structure. A paradigmatic example of this situation is the case of the Annex Building, where, in addition to the structure, the finishes only considered the placement of glazed facades, glazed skylights and the ceilings in suspended metal trellis. In the case of the exhibition halls, there was also the cladding of the vertical faces.

This aspect initially required that the Architect's concept for the project be internalised, which eventually happened naturally, because, from the beginning of the work, we found ourselves in tune on both the ideas and the expectations we all had for the project. Then, and even before the design work itself, the general criteria for the design of the buildings were easily settled together, and these were eventually strictly implemented during the various phases of the project. It appears now, after all the technical issues have been overcome and solved, that, in general, there are no deviations when we compare the photomontages prepared in the initial programme with the end result of the buildings as built, which admirably demonstrates the knowledge and experience provided, particularly by the architect, Paulo Mendes da Rocha, from the early sketches and from the first ideas outlined for the project.

**Fig. 21** 3D street view of the auditorium from Junqueira Street

**Fig. 22** Street view of the auditorium from Junqueira Street under structural construction phase

**Fig. 23** Street view if the main building under structural construction phase

## Exhibition Pavilion – Main Structure

The main body of the exhibition pavilion is a parallelepiped, 126 m long, 48 m wide and 12 m high, supported on 14 circular pillars 1.80 m in diameter and about 4.5 m high.

With these features, the choice of a metal structure was almost mandatory, since only by taking advantage of structural lightness and lightness of construction as a whole would it be possible not to penalise the 14 supporting pillars too much.

The structure is composed of four major triangulated main beams, about 12 m in height, arranged along the longitudinal walls of the exhibition hall naves, which ensures that all vertical loads applied to the building are transferred to the pillars.

In conjunction with the structural elements that support the elevated first floor and the roof, these beams establish a latticework of resistant perpendicular, interconnected planes, which ensure the lateral stability of the elements of each of these planes and appropriately transfer any horizontal forces due to earthquake or wind to the pillars and the central core.

The main beams in the intermediate alignments are 12 m apart from each other and supported on four pillars, forming 3 consecutive spans, each of 42 m. In turn, the beams of the alignments of the longitudinal façades are 18 m apart from the former, and provide three supports, forming two intermediate spans of 42 m each and two cantilevered spans at the ends of 21 m each. At both ends of these beams, there are two triangulated transverse beams of the same height, which close off the volume of the pavilion. The upper and lower chords for these triangulated beams are in the HEB550 profiles, which, in addition to the structural function, also have an architectural role once the end and side lattice girders mark the bottom and top of each of the façades.

The vertical and diagonal bracings are usually composed of H-type plated profiles, laid out with the highest inertia orientated along the plane of the triangulated beams. The vertical bracings located over the pillars are composed of HD400x744 profiles that, unlike all others, are orientated with the largest inertia perpendicular to the plane of the lattice girders.

The lower chord for the beams is interrupted in these vertical bracings, which are extended to the top of the pillars. For construction reasons, it was necessary to limit the width of the chords and diagonals to 310 mm, which obliged that the vertical bracings be made considerably slimmer. To minimize this effect, advantage was taken of the HEA100 profiles, the purlins supporting the façade claddings, linking them together via metal plates and giving them continuity in the area of the vertical bracings, which enabled a significant reduction in the lengths of buckling, thus taking more effective advantage of the section of the vertical bracings.

The vertical bracings of the main beams also ensure the transmission of loads coming from external elements that are connected to them, such as the profiles of the latticework beams of the floor, the metal beams of the intermediate floors, at level 2, the truss beams of the roof and the metal profiles of the structure of the outdoor veranda, which, on the southern façade, is supported by the structural beam of this alignment.

The joints between the main nodes of these beams are welded. The components are prepared in the factory and shipped to the site separately, and are then welded in situ on tables prepared for this purpose, so as to form parts that can be raised with the beams on site. Finally, to complete the structure, these members are placed in their final position and welded together, obviously without forgetting the transverse elements necessary to ensure stability of the whole during various stages of assembly.

**Fig. 24** View of the main building under structural construction phase from the second level of the auditorium

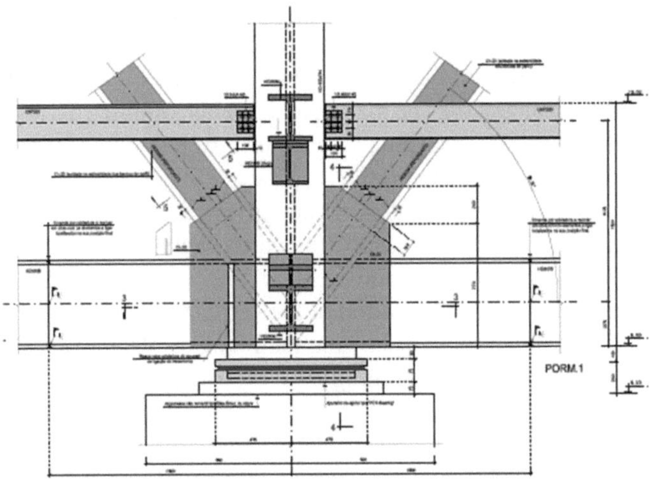


**Fig. 25** Detail of the structural support of the main building's external wall

**Fig. 26** Detail of the structural support of the main building's external wall under construction

**Fig. 27** Structural BIM model

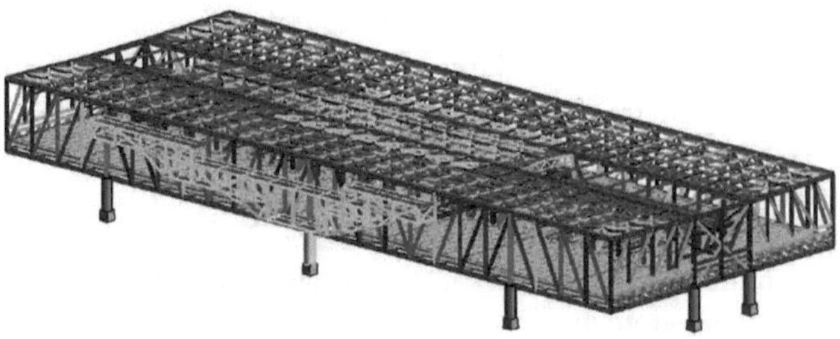

**Fig. 28** Roof structure under construction

## Exhibition Pavilion – First Floor Slab

The Level 1 slab, situated about 6.50 m above the ground, is supported by a system of triangulated cross-beams, 5.25 m apart, which receive the purlins, generally 2.25 m apart. The transversal beams are supported on the 4 large longitudinal beams, thus presenting 3 spans, one intermediate 12 m long, and two at the ends of 18 m each. These beams comprise chords, vertical and diagonal bracings in HEA and HEB profiles, and are continuous over intermediate supports. The end supports also use an extension of the lower chord of these lattice girders up to the support, taking advantage of the partial fixing that is created there to reduce the buckling lengths of the vertical bracings of the main beams.

In turn, for the purlins, it was decided that IPE140 profiles would be used, anchored at two intermediate points by an intermediate diagonal horizontal strut, thus ending up with 3 spans of 1.45 m, 1.45 m and 2.35 m. At the level of the lower flange of the transverse lattice girders, the struts are anchored by a set of 27 rods, which, in the case of asymmetric loads, ensure the reaction necessary to counteract the horizontal component of the force transmitted by the struts. At the level of the purlins, it is assumed that the upper flange will be anchored by the steel composite deck itself, and that the lower compressed flange, in the zone of the profile on the struts, will be anchored to slackening of the 10 rods to be welded directly onto the upper surface of the flange of the purlin profiles.

The Level 1 slab itself presents a set of particular characteristics that basically arise from the importance attached to it in solving a very diverse set of situations. In fact, in addition to its structural functions, the solution is required to meet the needs of thermal insulation, provide the final finishing for the museum's floors and allow the embedding of the pipes that circulate the radiant liquid for the air conditioning of the spaces.

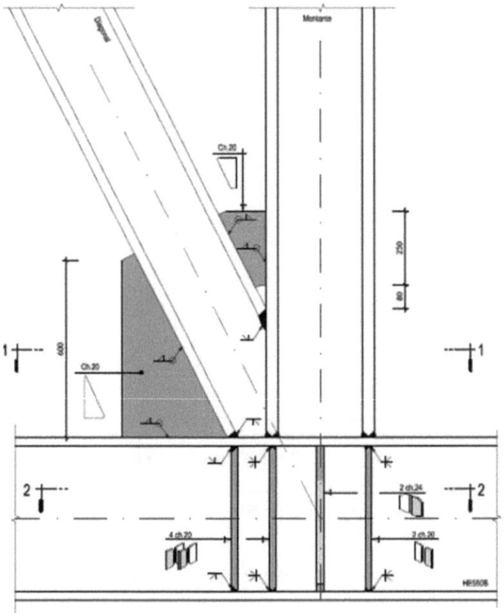

**Fig. 29** Detail of the beam of the main building's external wall

**Fig. 30** Detail of the beam of the main building's external wall under construction

One of the fundamental premises established from the outset was not to consider expansion or construction joints along the entire area of the exhibition halls.

In terms of determining the finish, various design solutions were tested during the design process, constructing samples of concrete polished with different aggregates and compositions. The option eventually chosen was for a floor with white surface hardener, smoothed onto the solid slab in white concrete, which, in addition to meeting aesthetic requirements, offers an excellent performance as a flooring material, especially in terms of resistance to impact and wear.

In the construction phase, and for a variety of reasons, it was later decided to make a grey concrete slab with a surface hardener and final finish obtained by applying a lithium-based sealant and subsequent polishing. With this solution, a better surface strength and tightness is obtained and, moreover, the final polishing can be done at the end of construction, thus enabling the inevitable marks and damage the floor is subjected to during the construction period to be eliminated or at least minimised.

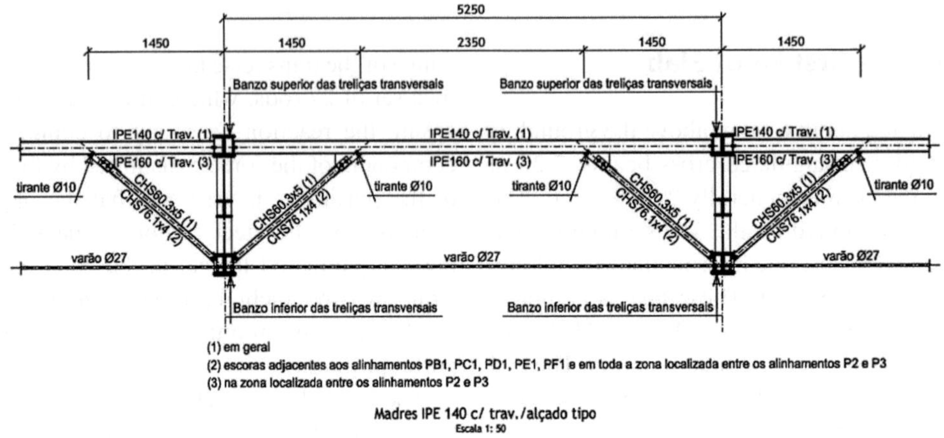


**Fig. 31** Detail of the roof truss structure's main building

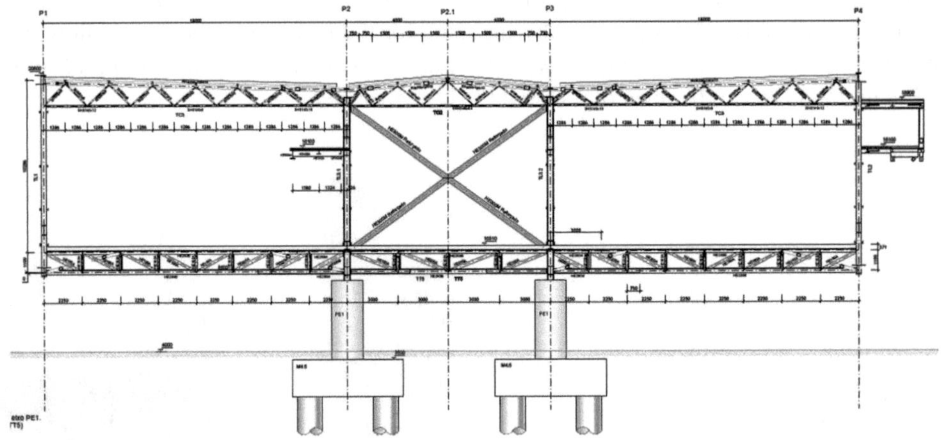

**Fig. 32** Cross section of the roof truss structure's main building

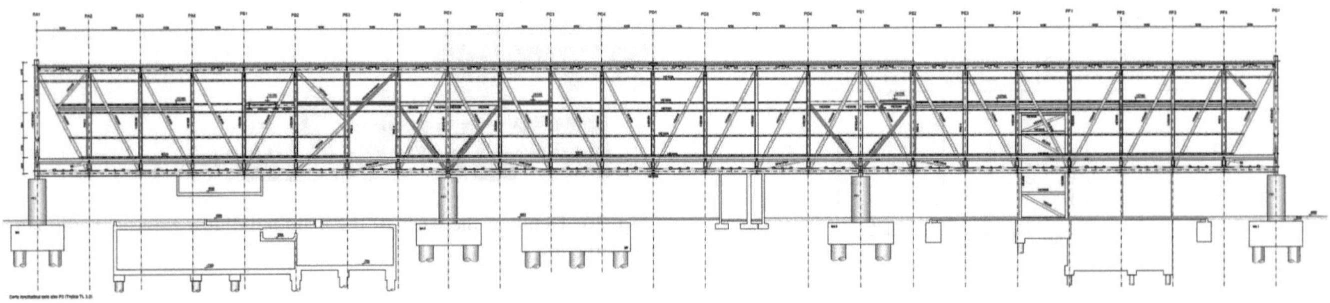

**Fig. 33** Structure of the main building, long section

The constitution of this slab includes a profiled steel sheet, 1.5 mm thick, supported on the purlins, receiving the filling of the conduits with lightweight expanded clay concrete, a bond-breaking film, an 8 cm thick layer of expanded polystyrene thermal insulation and another bond-breaking film, over which concrete is poured onto the 15 cm thick concrete slab.

The radiant floor tubing is embedded in the slab. For the absorption of horizontal loads into the floor, the proposal is for it to be connected to the central core of stairs, as well as to the main metallic structure, in the middle third of the slab, over a length of 42 m. The freeing of the slab in the two end thirds results from the need to reduce the stresses due to shrinkage and thermal variations. For seismic action in the transverse direction, the end thirds of the slab work as a cantilever, ensuring, through a diaphragm effect, the transmission of horizontal actions to the fixed supports of the intermediate pillars and to the reinforced concrete core of stair E2.

**Fig. 34** Internal view of the main building during the structural construction phase before the concreting

**Fig. 35** View of the main building slab from the lower level

To limit and control cracking, it is expected that longitudinal and transversal pre-stressing will be applied, sized to ensure a 1 MPa minimal residual compression for most situations. The pre-stress is adherent, with mono strands of "0.6" arranged in centred metal sheaths, spaced 0.40 m between them.

In dynamic terms, the slab for Level 1 of the exhibition hall has its first natural frequencies between 3 and 3.50 Hz, therefore within the critical range that would make the slab susceptible to producing uncomfortable levels of vibration.

Taking this into account, several studies were carried out according to the latest publications on the subject, and it was concluded that, although the natural frequencies lie within the critical range, the masses involved have very significant values that, therefore, ensure suitable comfort conditions.

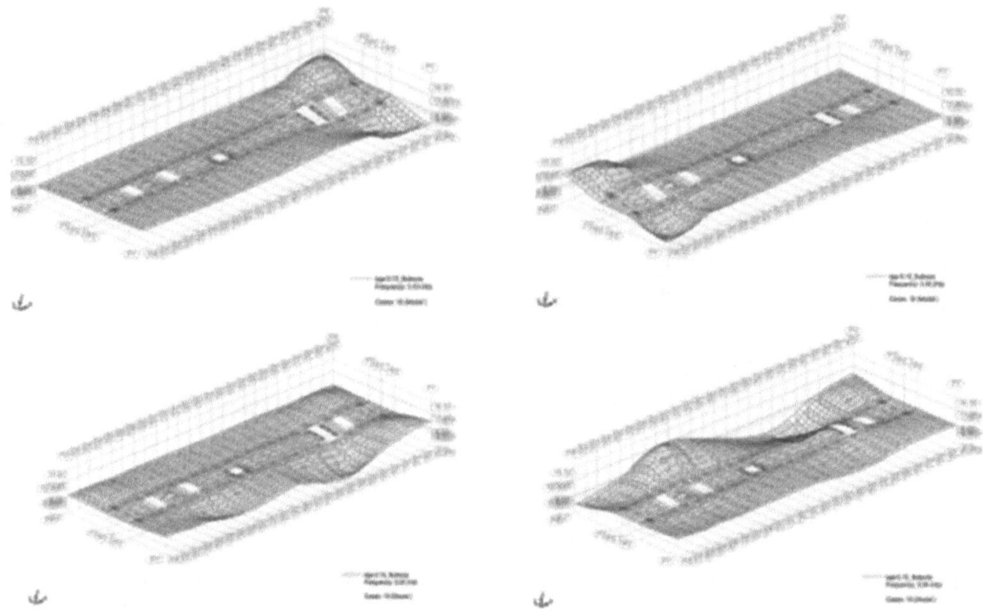

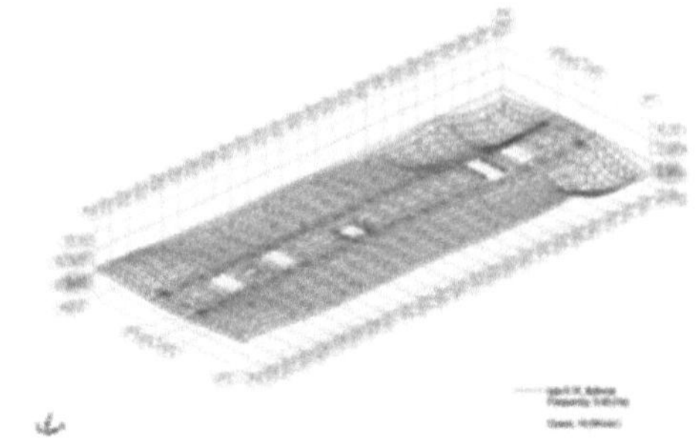

| Modo | Freq (Hz) |
|---|---|
| 7 | 3.09 |
| 8 | 3.16 |
| 9 | 3.23 |
| 11 | 3.34 |
| 12 | 3.38 |
| 14 | 3.46 |

**Fig. 36** Simulation of the main building slab structural efforts

Fig. 37 Main wall building under construction

Fig. 38 Main wall interior under construction before the concreting phase

Fig. 39 Main building's external wall under construction

Fig. 40 Main building's external wall module under construction

## Exhibition Pavilion – Level 2 and Roof

The slabs for Level 2 are only in the central nave, between the two large longitudinal intermediate beams, and are composed of composite slabs that are 12 cm thick in the general areas, while in the plant rooms, partially to cope with the weight of the equipment housed within them, but also, particularly, to minimise and control the transmission of vibrations emitted by this machinery to the main structure, it was decided to use 20 cm thick slabs.

As in the exhibition halls, here, the finish is also applied directly onto the concrete floors, initially through mechanical smoothing and then by polishing after prior application of a lithium-based sealant.

In this zone, the supporting metal structure consists of the main metal beams spaced at 5.25 m, which bridge the 12 m between the large longitudinal beams, supplemented by secondary profiles at right angles, where they support the floor slabs.

At this level, metal walkways are also provided that traverse the naves of the exhibition halls, making connections to the annex to the north and the external veranda to the south. These walkways are composed of pairs of HEB550 beams that take advantage of their mixed behaviour to bridge a total span of 18 m.

## Exhibition Pavilion – Foundations

For the main pillars and cores in reinforced concrete, a solution of indirect foundations was adopted, using reinforced concrete piles cast in the ground and inserted into the volcanic complex with a W4-3 weathering degree and a F5-4 fracturing degree; the length of insertion varies depending on the loads to be transmitted to the foundation.

The actual length of the piles is between 6 m and 15 m, their diameters varying between 600 mm and 1500 mm. The design considered a permissible stress at the tip of the piles of 4000 kPa and tangential stress values along the length of the insertion of 75 kPa.

## Annex Building

The annex comprises the construction of two structurally independent bodies, the auditorium and the building that houses the administrative services and the restaurant.

## Annex – Main Structure

In this last volume, the main structure consists of a grid of four reinforced concrete frames at right angles to each other, and with axes formed according to the sides of a square, 45 m by 45 m, complemented by two large reinforced concrete shafts housing the vertical accesses by elevators and stairs.

The north and south frames are identical and have two uprights each, near their ends, with a thickness equal to the entire frame, 0.80 m, and variable dimensions in the plane of the frame between 2.00 m along the foundation, increasing to a maximum of 8.00 m in the area of the connection to the intermediate horizontal strut, then reducing to a minimum of 0.80 m at the top. The intermediate horizontal strut is located at about half the height, with a rectangular section of 0.80 m x 1.80 m that connects the two pillars of each frame.

Perpendicular to these and supported on the same uprights are the north and south porticos, consisting of pre-stressed reinforced concrete beams, 0.80 m wide and 5.60 m high.

It is by way of these porticos that the main vertical loads are transmitted from the building to the foundations. Furthermore, and although the reinforced concrete shafts absorb a large part of the horizontal loads due to earthquake and wind, the peripheral arrangement of these porticos also enables them to play a key role in controlling the rotational movements of the building.

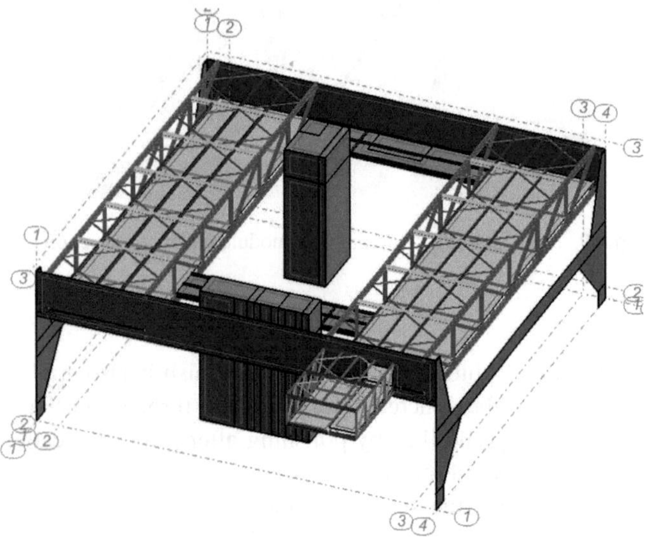

**Fig. 41** BIM model of the auditorium's structure

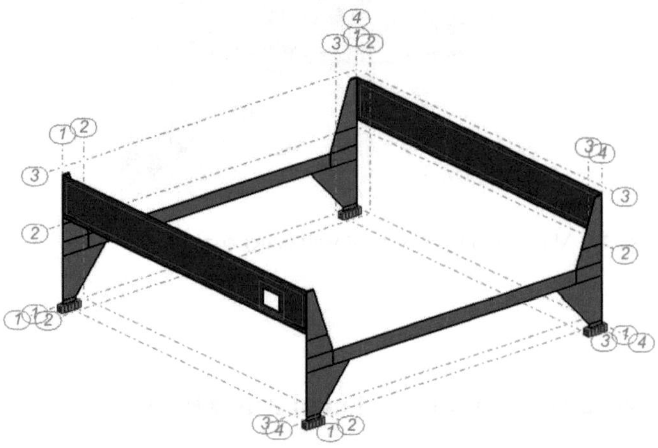

**Fig. 42** BIM model of the auditorium's concrete structure

**Fig. 43** Auditorium structure under construction

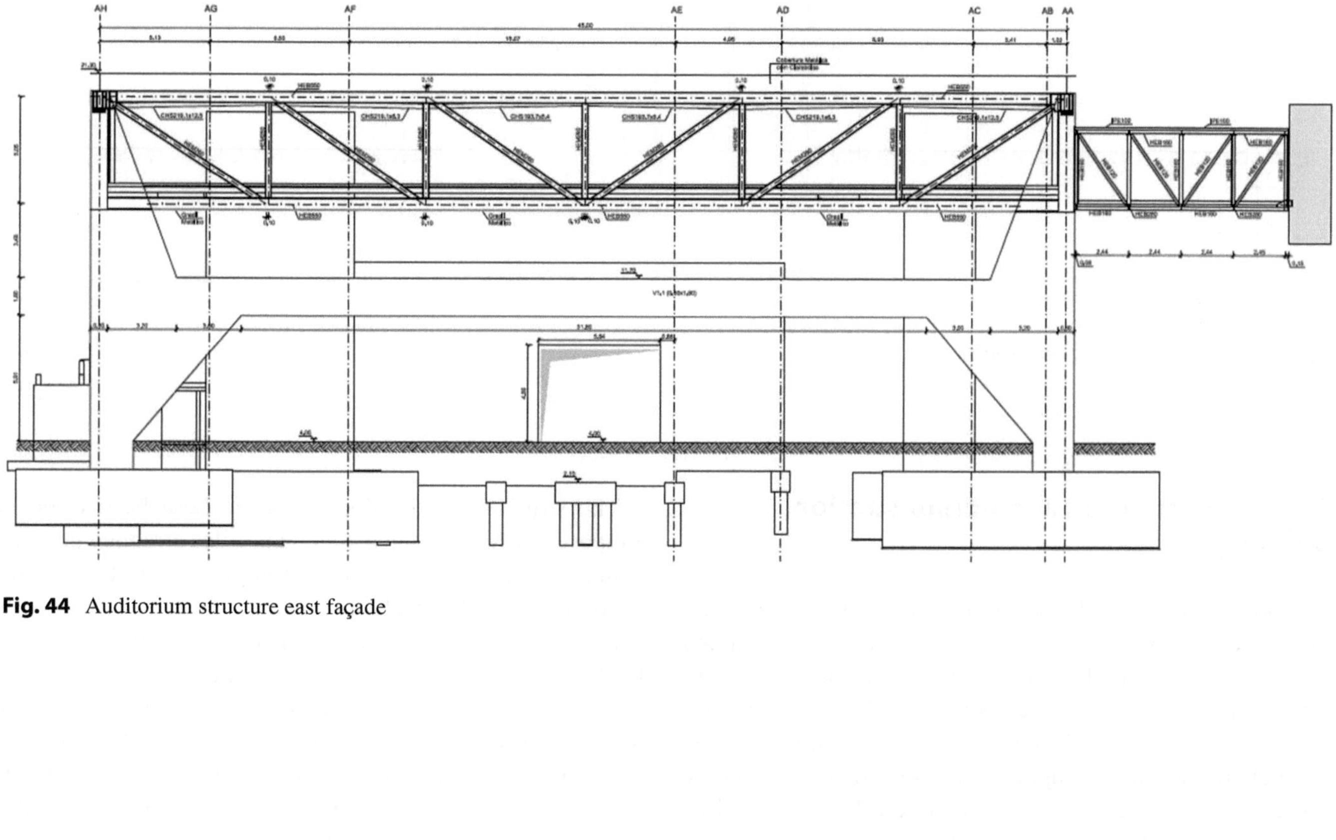


**Fig. 44** Auditorium structure east façade

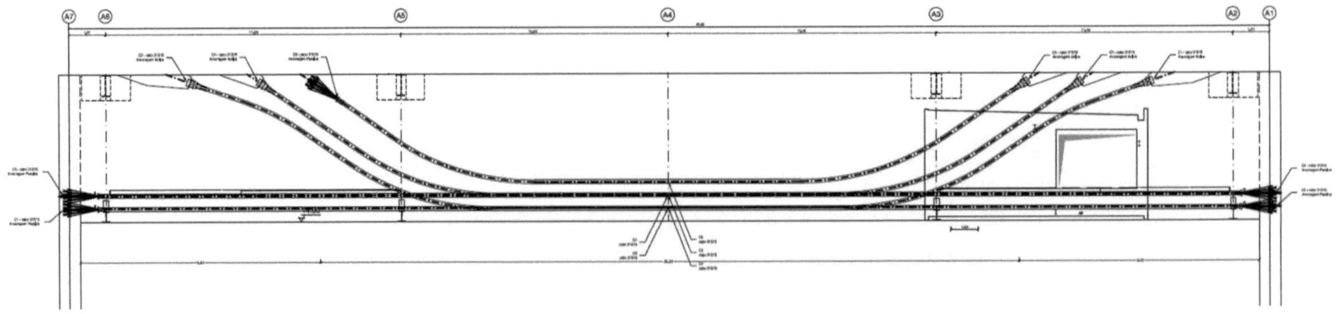

**Fig. 45** Detail of the beam structure of the auditorium

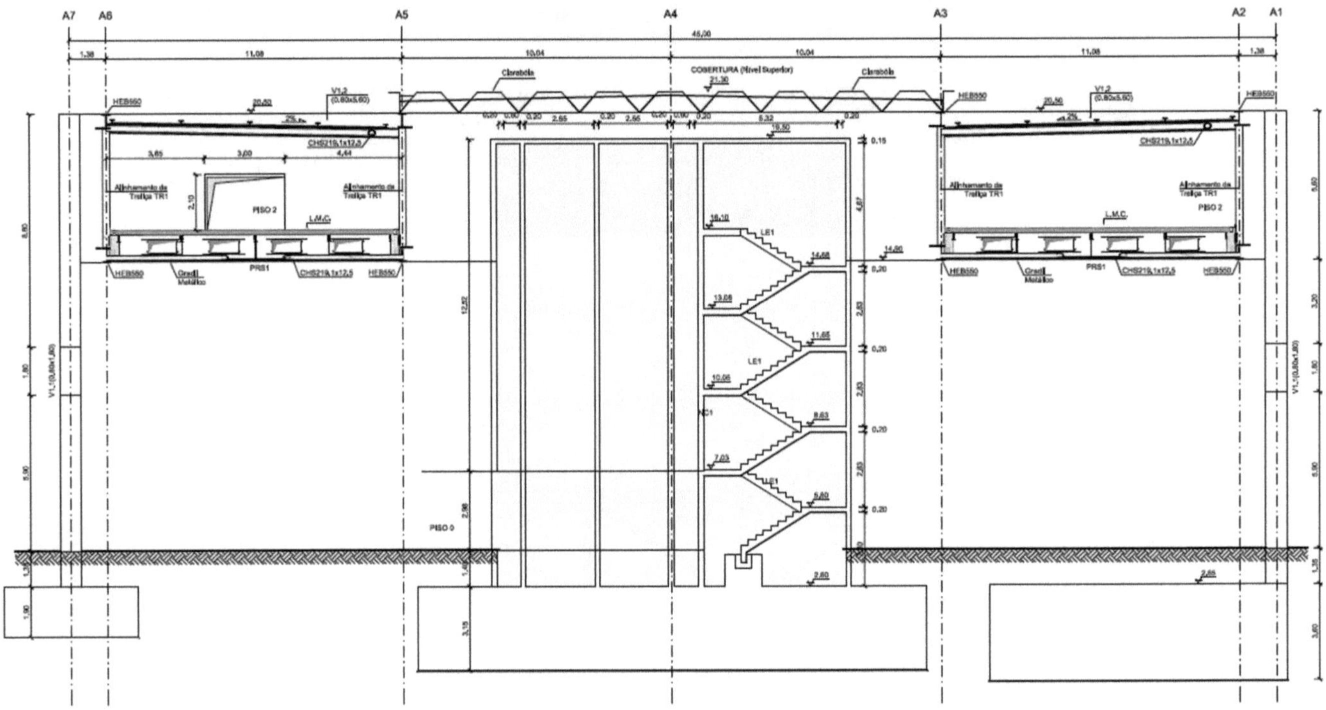


**Fig. 46** Auditorium structure cross section

## Annex – Restaurant and Administration Building

The areas that house the restaurant and administration services are on only one level and the respective roof area and are a suspended parallelepiped volume, 45 m long, 11 m wide and about 5 m high. At their ends, these volumes are supported on the deep beams that form the east and west concrete frames, thus bridging a total span of approximately 45 m. The main structure of each of these volumes comprises two large parallel triangulated steel beams, 5.50 m high and at intervals of about 11 m.

The structure of the floor rests on these beams and is composed of reconstituted welded profiles arranged perpendicularly to the main beams and aligned according to the position of the vertical bracings of these beams, which receive the longitudinal purlins that support the composite floor slab. The roof structure is identical to the floor, while the profiles that constitute it are naturally lighter, because it only has a sheet metal coating. In addition to the structure that assures the transmission of gravity loads to the main beams, horizontal bracings are also provided throughout the floor and roof structure, ensuring the lateral bracing of the main beams.

Each of these two suspended volumes rests on the two large pre-stressed deep beams at only four points, arranged at the ends of the large triangulated beams. These supports are located at the level of the upper chord and employ rigidly controlled neoprene structural bearings. All of the vertical loads are transmitted to these supports. Horizontal actions, due to earthquake activity primarily at the ground level, in a longitudinal direction, are considered to be transmitted in part by the structural bearings and in part by a system of shear lugs provided in the lower chord of the main interior beams.

This system of shear lugs, on the one hand, enables seismic actions to be transmitted by contact to the north concrete frame or to the south concrete frame, depending on the direction of the actions, and on the other, prevents embedding of the large triangulated beams into the reinforced concrete frame. In the transversal direction, the horizontal loads acting at the floor level are transmitted through the same buffer system, as well as through the direct connection planned between the concrete slabs and the reinforced concrete porticos. To avoid undesirable concentration of stresses in the most rigid areas near the uprights, this direct connection of the slab to the porticos is only envisaged for the lower half of the floors.

It should also be noted that the rigidity of the neoprene supports enables them to ensure, for quick actions such as seismic activity, the transmission of horizontal loads by distortion, and for slow actions such as temperature variations, that forces are not prevented, which allows the distortion of the structural bearing without the transmission of loads to the supporting element.

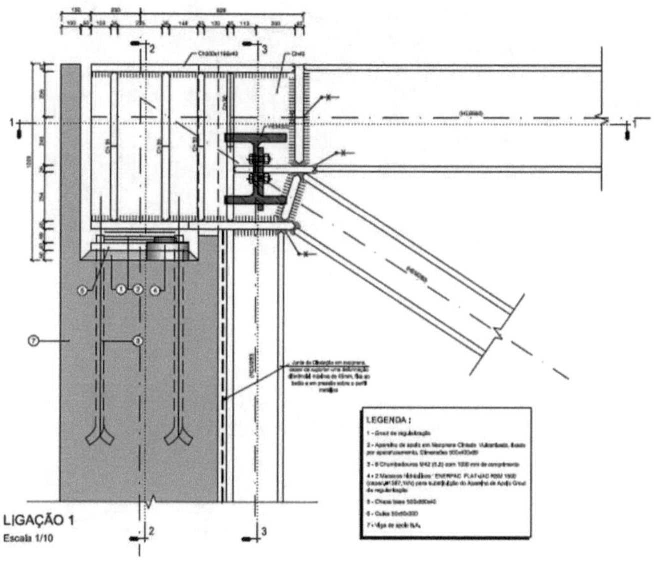


**Fig. 47** Detail of the connection of the steel structure with the concrete wall

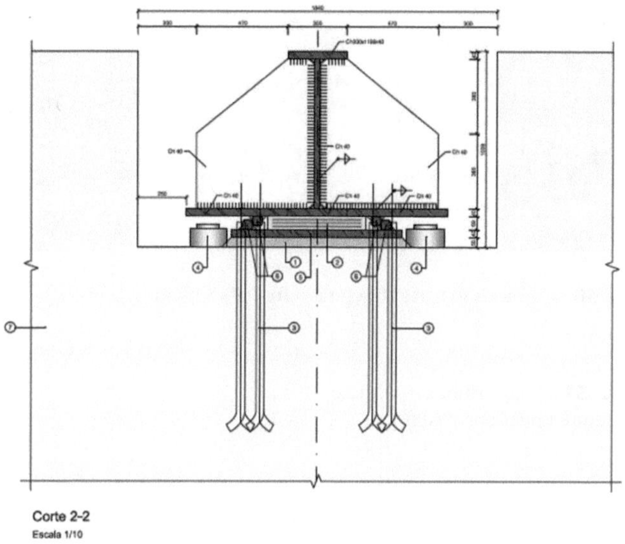


**Fig. 48** Detail of the connection of the steel structure with the concrete wall - cross section

**Fig. 49** Detail of the connection of the steel structure with the concrete wall under construction

**Fig. 50** Auditorium truss structure before placement

**Fig. 51** Auditorium's concrete structure under construction

**Fig. 52** Auditorium under construction before the finishing phase

**Fig. 53** Auditorium structure finalized

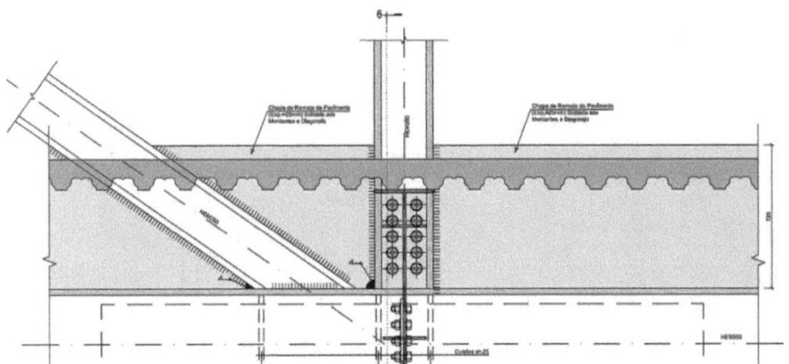


**Fig. 54** Detail of the connection between the slab and the steel structure, long section

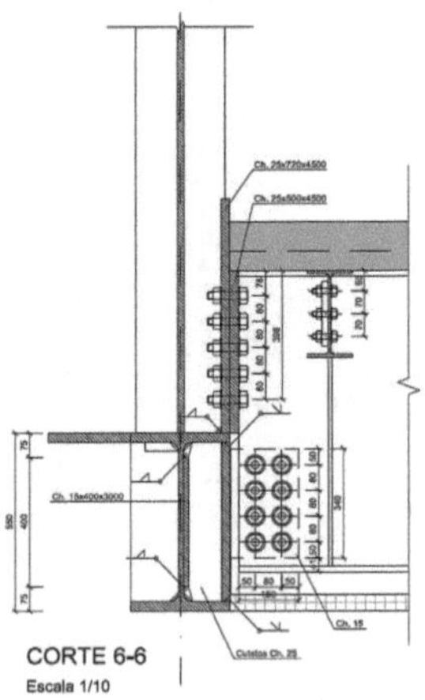


**Fig. 55** Detail of the connection between the slab and the steel structure, cross section

**Fig. 56** Auditorium skylight's

## Annex – Skylight

Between the restaurant and the administration bodies, a metal sheet roof is envisaged, made of a set of parallel V-shaped beams with a maximum width of 1.20 m and about 0.80 m in height, made of a 6 mm sheet. Near the "V" top end, another 6 mm sheet is envisaged, welded to the "V" wings, positioned so as to provide a 2% rake that ensures rainwater drainage. The axes of these beams are set about 2.25 m apart, with further sheets between them, arranged at right angles and replicating the same type of V-shaped beam. This set of right-angled beams enables a number of skylights to be created with a very interesting visual appearance.

**Fig. 57** Auditorium's skylight in position before assembly

**Fig. 58** Auditorium's skylight before assembly

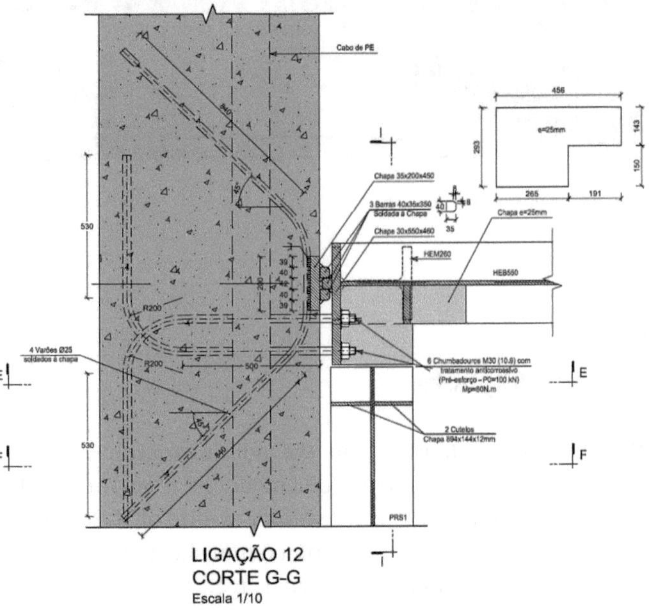


**Fig. 59** Auditorium skylight's detail

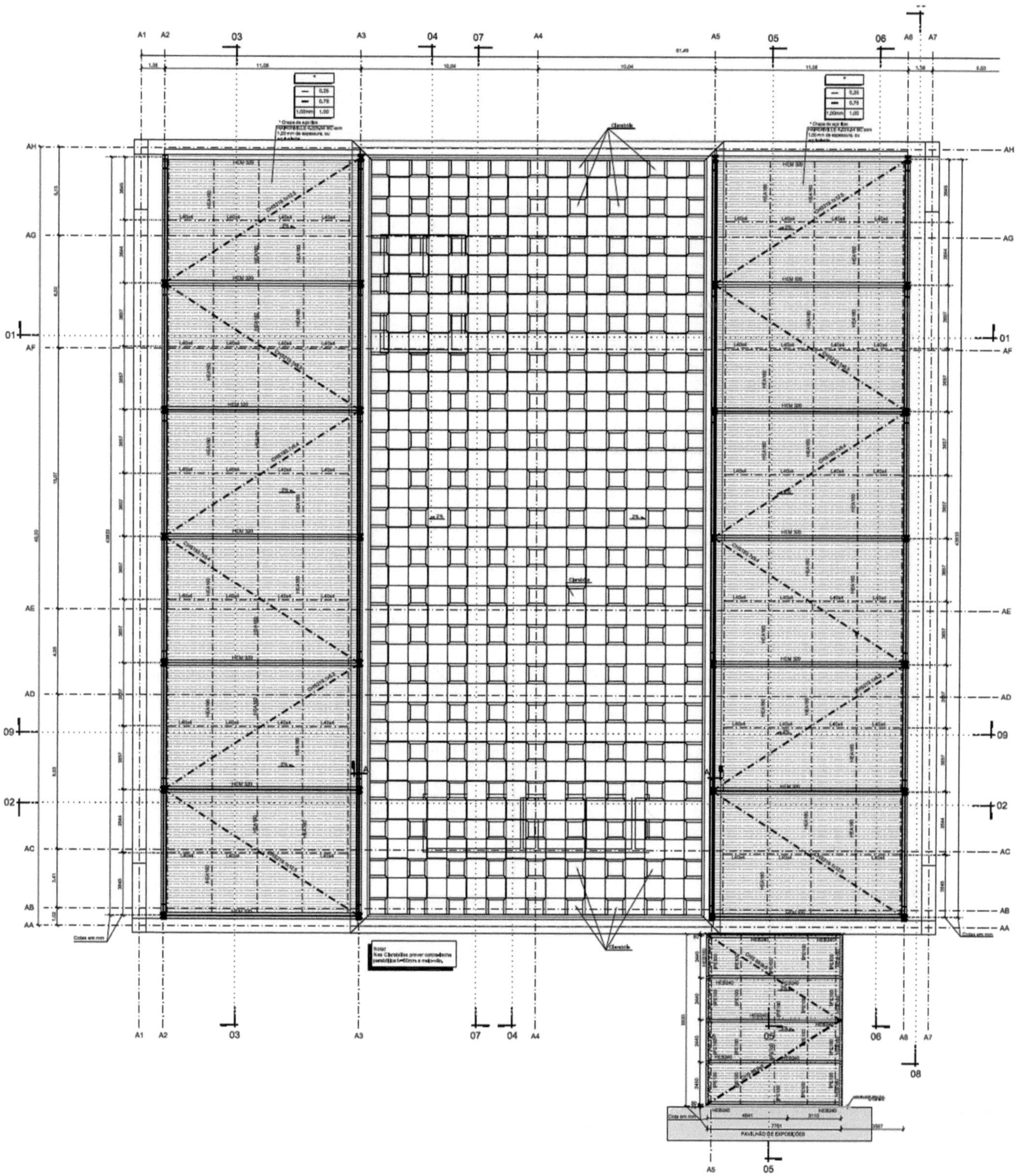


**Fig. 60** Auditorium roof plan

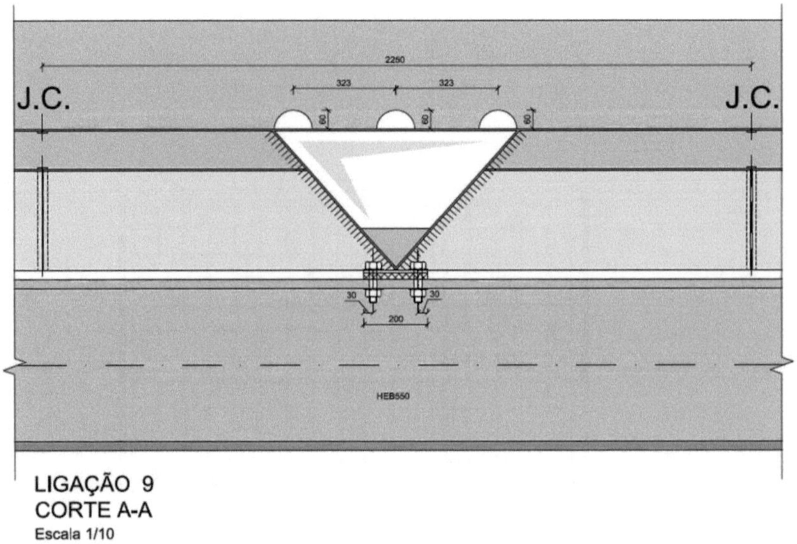


**Fig. 61** Auditorium skylight's connection detail long section

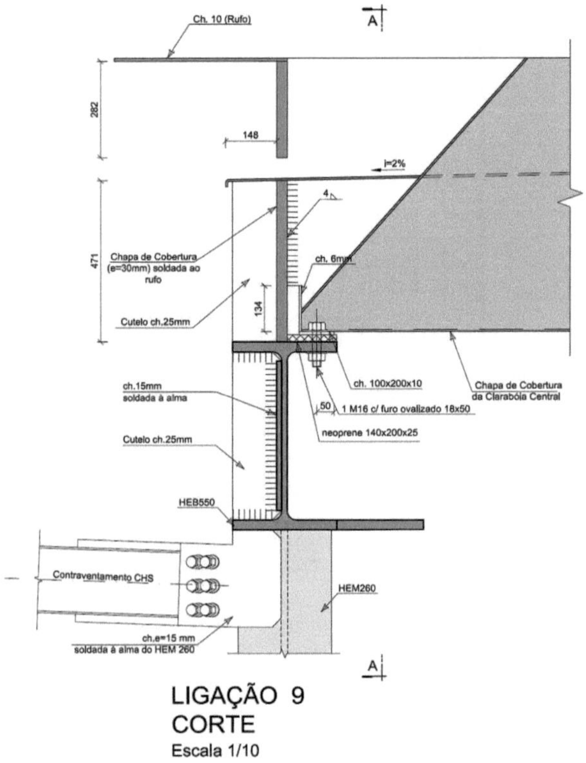


**Fig. 62** Auditorium skylight's connection detail cross section

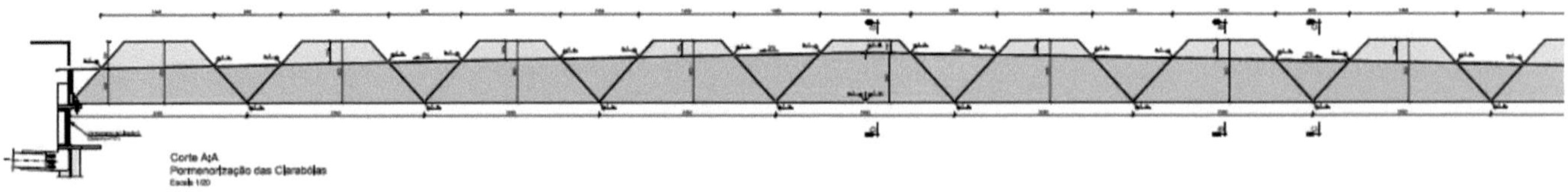

**Fig. 63** Auditorium skylight's long section

**Fig. 64** Auditorium skylight's under construction

## Pedestrian Walkway

The pedestrian walkway comprises 3 structurally independent bodies. To the north, still within the confines of the annex, 3 ramp sections provide access to the deck. The first ramp section is in reinforced concrete with deactivated finishing, while the others are a metal structure bridging a total span of about 37 m, their ends resting on cantilevers embedded into the concrete pillars, which are placed in the gap between the parallel ramp sections. In these sections, the deck's transversal section comprises 4 girders in HEA600 profiles, coupled at points by means of crossbars set 1.50 m apart. The transversal rigidity of the deck is ensured by a cross-triangulation system with L90 x 90 x 9 angles. The cladding is made of extended steel plate.

For the transition to the deck, a support with an expansion joint is envisaged, which releases the longitudinal displacements of the deck.

The deck has a cross-section of the same type, but with only 3 longitudinal beams in HEB600 profiles, and covers 7 consecutive spans with total maximum lengths of about 30 m. The support is made of 3.00 m x 0.25 m reinforced concrete walls with which it is embedded in the foundation.

Access through ramps is complemented by direct access from the deck, through stairs to the railway platform and the north pavement of Avenida da India, and by lift or stairs to the south pavement of Avenida de Brasília.

The deck's longitudinal bracing is essentially ensured by the structure that makes up the stair and elevator shaft frame, where the installation of a vertical bracing system is envisaged, made of triangulated bars. The cross-bracing of the deck is ensured by the rigidity that the concrete walls on which it rests have in that direction.

As in what happens near the north abutment, the walkway ends to the south through the unfolding of the access ramps into parallel sections. Here as well, the starting section is made of deactivated reinforced concrete, the others being of a metal structure with the same cross-section as the deck.

To the north of Avenida da Índia, the access stairway, to be made of exposed white concrete, displays a spiral shape, the connection to the deck being effected through plates and rock bolts embedded into the concrete. This connection does not allow for relative displacements between the stairs and the deck, so the staircase is designed considering the corresponding differential settlements in the longitudinal and transversal directions.

For aesthetical reasons, the option of a deck of marked slenderness was chosen, which obliged consideration of tunned mass damper (TMD's) devices to correct fundamental vibrations in the structure. These devices are installed under the deck, between the girders, and will be calibrated on the basis of confirmed dynamic studies to be carried out after the walkway is constructed.

For the rail guards, the choice is pre-fabricated panels of white concrete, 80 mm thick and with maximum lengths of about 2.70 m, fitted into the metal uprights, which are first welded to the deck's girders.

As in the buildings, an indirect foundation solution is also adopted here, generally through reinforced concrete piles, moulded on site, embedded into the basalt rock mass, with a weathering degree of W4-3 and F5-4. In the area of the railway platform, the restrictions of available space impose the use of an indirect foundation solution, but using micro-piles.

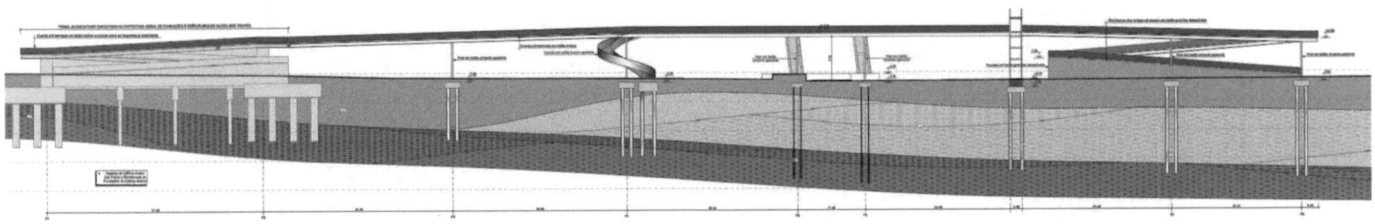

**Fig. 65** Walkway structural long section

**Fig. 66** Walkway structural 3D mode

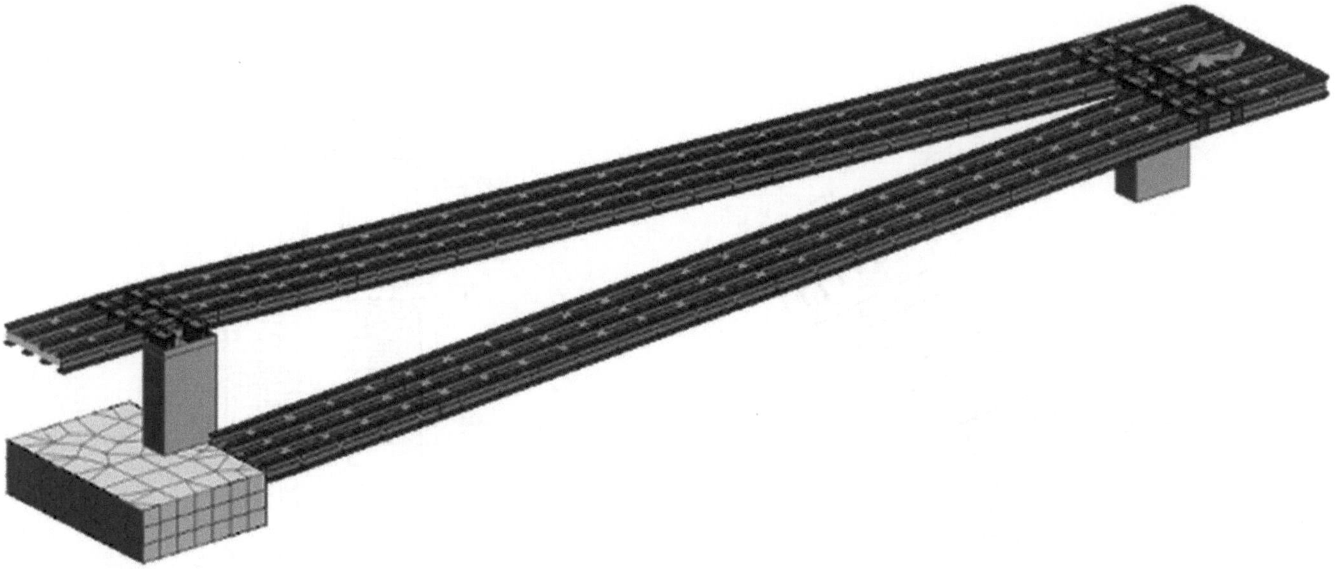

**Fig. 67** Walkway ramp structural 3D model

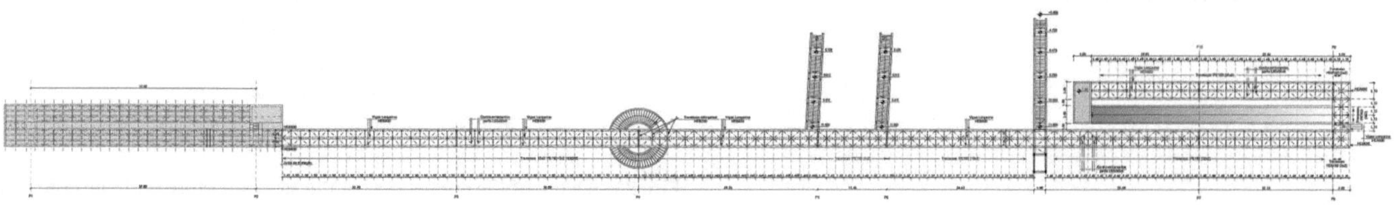

**Fig. 68** Walkway plan

**Fig. 69** Walkway under construction

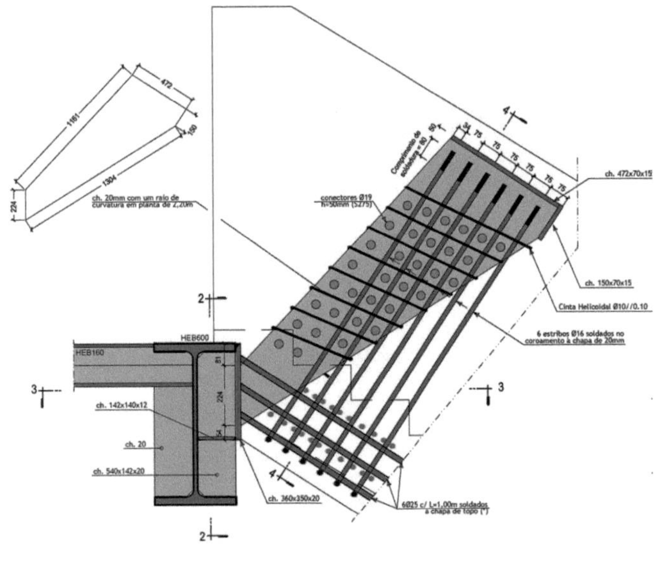


**Fig. 70** Walkway connection details

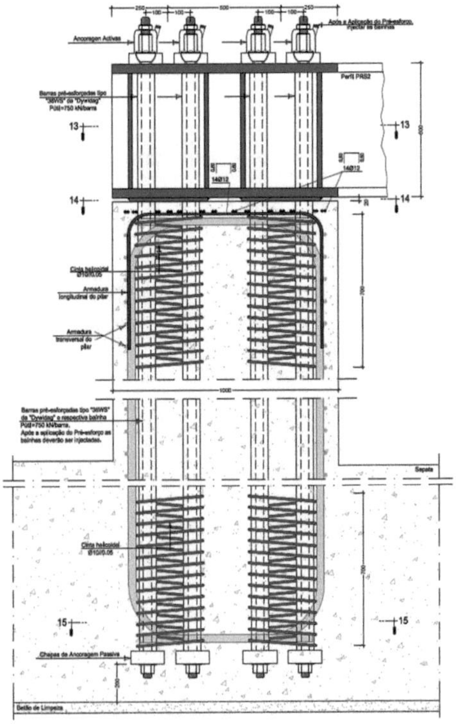

**Fig. 71** Detail of the connection of the steel structure with the concrete wall - cross section

**Fig. 72** Walkway ramp under construction

**Fig. 73** Walkway under construction in the connection phase

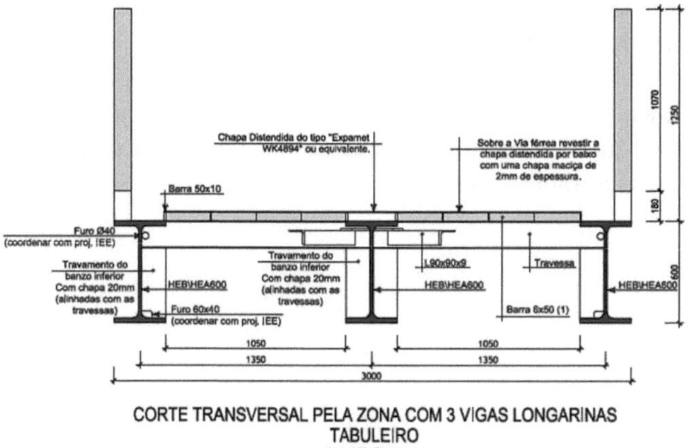


**Fig. 74** Walk way structure

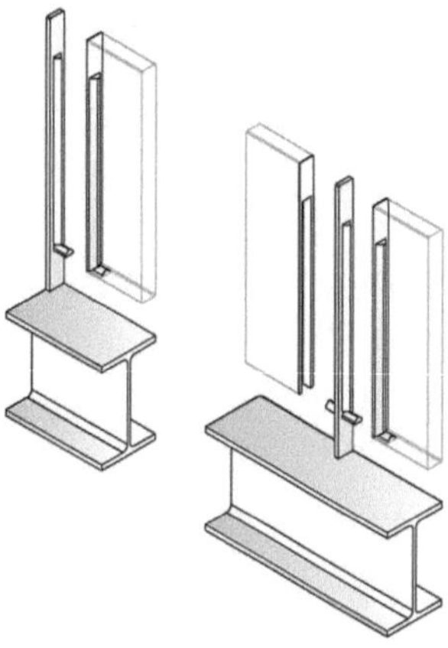

**Fig. 75** Walkway handrail detail

## Hydraulic Installations

Hydraulic installations and equipment for the building of the new Coach Museum were designed so as to fulfil functional needs, complying with comfort, reliability and safety requirements, and considering, whenever possible, options that enhance the sustainability of the new construction.

## Water

With regard to water management, for a more efficient and judicious use of this resource, the option was for a mixed supply solution, making use of drinking water only for drinking and applying a system that harnesses rainwater to supply all of the equipment that requires water from other sources.

Toilets, urinals and irrigation of outdoor spaces will be primarily supplied from the rainwater reserve. This tank was built in the basement in order to protect the water stored there from thermal variations and light. The design includes a first-flush system to detour the first water, because, as is known, the initial minutes of rain contain a more significant pollution charge, which is not desirable to have in the tank. Upstream of the water distribution to toilets and urinals, in the supply line between the tanks of raw and treated water, the installation of a filtering and purification system is envisaged to avoid the risk of transmission of any contaminants into the water distribution system.

Rainwater used in the harnessing system is collected on the building roofs and channeled to the tank through a siphonic network (Pluvia da Geberit), water from the outside pavements being excluded for this purpose, since the level of contamination makes its re-use inadvisable. The whole system has been conceived so as to enable gravity drainage, directly to the public network, from the rainwater storage tank, whenever it is full.

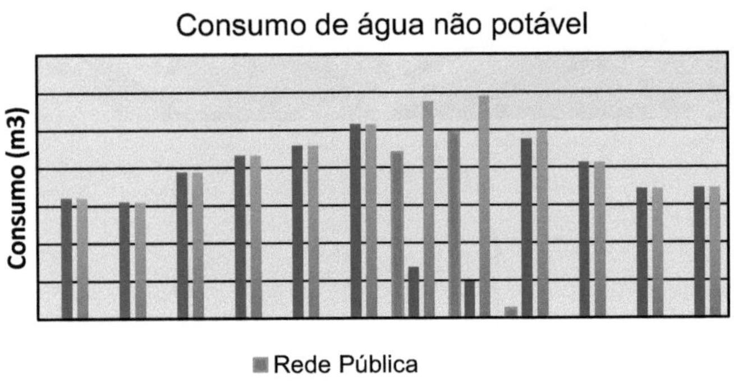


**Fig. 76** Graphic of the water consumption prediction

With the harnessing system, the need for a water supply from EPAL only occurs from July to September, corresponding to 18% of the total consumption.

## Sprinklers and Water Curtains

With regard to safety, notably to extinguish fires by means of water, the building includes, in addition to a fire-fighting system (RIA), a sprinkler and water curtain system, supported by a 260 $m^3$ water reserve and two independent pumping groups (one for the RIA and the other for the sprinklers and the water curtains).

In the exhibition hall, at the ground level, comprising the workshops and the other supporting areas, a wet-type sprinkler system is envisaged. In this type of installation, the system will work under load, so the system is activated as soon as the sprinkler's bulb reaches the predefined temperature.

In the museum areas, owing to the specificity and value of the exhibits, the option was for the installation of a double interlock pre-action sprinkler system. In this way, it is possible to avoid accidental damage from water, both in the piping system and in the sprinklers. Two events are needed for a delivery of water to occur: an alarm signal from the fire detectors and the operation of a sprinkler. If the alarm is activated but the sprinkler does not enter into operation, the water is not supplied to the piping system, nor does it reach the sprinklers. If a sprinkler enters into operation, but there is no alarm, the water also does not enter the system.

On level 2 of the annex, the installation of a water curtain-type instant flooding system is also envisaged, consisting of an open sprinkler network capable of causing the façade structure to cool down, since, for architectural reasons, it has no other fire protection system.

This system comprises a section of metal piping, laid down on the inside of the top of the structure, where the full cone nozzles required to ensure the formation of an even water curtain are placed.

**Fig. 77** Infrastructures construction phase

**Fig. 78** Springlers for fire detection

## Electrical, Telecommunications and Active Safety Installations

The approach to the design of the electrical, telecommunications and safety installations and equipment fits into the level of excellence defined for the building, and so technical accuracy, innovation, system optimisation and, in particular, their integration into the architecture were the main constraints.

It should be noted that it was an architectural choice to expose the various elements of the technical installations, and this required special care in developing the layouts and designing the systems so as to achieve a pleasant and thoughtful end result. In view of the amount of networks envisaged for the spaces in general, a colour code was established to facilitate identification of the various networks, referring to the specific function of each:

- Orange: Electrical installations and equipment;
- White: Mechanical installations and equipment;
- Red: Fire-fighting systems;
- Black: Hydraulic installations.

**Fig. 79** Temperature simulation

**Fig. 80** Infrastructure support

**Fig. 81** Infrastructure support with the different equipments

In general, the underside of the suspended ceilings is finished with a metal grid, with a 10 x 10 cm mesh, mainly intended to demarcate the space for infrastructure, as well as to achieve a sense of continuity. Bearing this in mind, the overall lighting solution considered the installation, over the industrial-type lighting grid, of a polycarbonate housing, which, in addition to giving it a high IP, also reduces the frequency of maintenance, with a high performance reflector equipped with T5-HO bulbs (High Output) and electronic ballasts.

In the exhibition halls, given that the coaches are highly sensitive to UV radiation, an overall lighting was calculated for the spaces of about 100–150 lux, reinforced at points by spotlights equipped with appropriate filters and fitted with the Globe system, with an adjustable focus, which enables forms to be defined with high precision at considerable distances.

To suspend the spotlights, as well as all other equipment that, owing to conceptual requirements, cannot be placed above the grid (such as speakers, video projectors, etc.), a detachable suspension system was created, which can thus be easily relocated to any of the grid "squares."

To reconcile constructive issues and the required widths for installing ventilation ducts and other infrastructure, the choice was made to use very thick false interior walls along which the various technical systems required are distributed. These walls are used to house larger equipment, such as electrical cabinets and some local supply and control equipment, as well as the museum's display cases, which are lit internally by fluorescent tubes and directional LEDs.

**Fig. 82** Simulation of the showcases

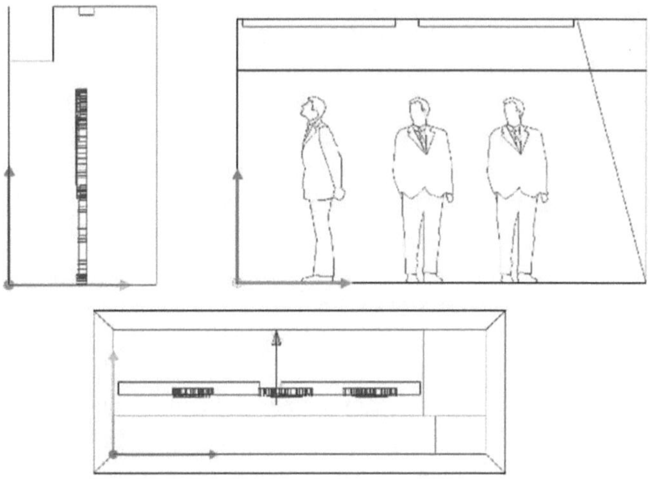

**Fig. 83** Showcases drawings

**Fig. 84** Showcase under construction

The design of the elevators, with a special focus on the two main lifts, capable of carrying 65 people at a time, was created with the idea that they represent the "21st century coach," the ride that offers a first look at the museum, through the openings in the cabin that provide a view to the inside of the "suspended ceiling."

With regard to telecommunications, a category 6 UTP cable and fibre optic structured network was created, covering the whole space, enabling changes to be implemented at any time in a simple way, such as a proposal for museum content to be shown through various multimedia equipment, whose location is easily adaptable, since they rely on solutions centralised at the servers.

Regarding security, in view of the valuable collection on display, as well as the experience of the existing museum, a configurable CCTV system was implemented, enabling the functions of control of people and intruders to be combined, as well as the continuous recording of images from the various spaces. In the exhibition halls, the DOME-type CCTV cameras, with HD varifocal lenses, allow for the definition of a perimeter around a certain object (a coach, for instance), proposed to coincide with the lettering elements of the museum project, in which, should a violation of this virtual barrier occur, an automatic alarm signal is triggered at the main security station. This system will reduce the need for local surveillance, in addition to optimising remote surveillance.

## Mechanical Installations

The current National Coach Museum in Lisbon "preserves today one of the most important and valuable collections of its kind in the world" (quoting Simonetta Luz Afonso).

The purpose of a museum is to preserve artefacts, materials and historical information for future generations, but also to provide access to this heritage for present generations, and to make it attractive.

Air conditioning systems take on an important role in achieving these aims, since they can contribute to minimising the deterioration of the collections, although at the other end of the spectrum, if incorrectly configured with regard to temperature and air humidity, they can actually accelerate this deterioration.

Different types of material require different optimum levels of humidity for their preservation. Collections that include different types of artefact or material require, therefore, a compromise as to the value of air humidity to be provided. On the other hand, in the case of collections that have never been in a very controlled environment, tight control of these parameters may actually be counterproductive.

The project for the Mechanical Installations and Equipment (HVAC) exists to propose solutions that will provide the museum with conditions of comfort for various situations of usage of the collection, but also the thermal and hygrometric conditions required for the preservation of said collection, in light of a compromise between preservation and human comfort in the exhibition, storage and restoration spaces and thermal comfort in the auditorium and the areas of access to the museum, as well as ensuring normal and emergency ventilation for those spaces that so require it and the production of hot sanitary water, according to the needs defined. All of this is always from the point of view of energy optimisation and associated operational costs.

**Fig. 85** Main room in the finishings construction phase

## HVAC Systems

The design of the environmental treatment systems seeks to meet the diversity of thermal demands, brought about by the functions of the different areas that make up the building, notably the exhibition and storage spaces, the restoration rooms/workshops, the auditorium, the administrative offices and the restaurant spaces.

Humidity control is paramount in museums and is especially aimed at ensuring that materials:

- do not absorb or promote water condensation, which might accelerate adverse chemical reactions or microbiological attacks;
- do not become dehydrated, losing their mechanical resistance and/or flexibility.

The degree of deterioration of the materials is not only a result of the level of humidity, but also the frequency and cycle of its variation.

Since the ideal values for the temperature and humidity parameters are different, and not always consensual, according to the materials to be preserved, it is fundamental that the client and/or the collection curator have their say on the values of temperature/humidity interaction that are acceptable for the exhibition, restoration and storage areas of the Coach Museum and on the different solutions proposed and compromises adopted, since this is essential to validate the costs of air conditioning systems, as well as the costs associated with architecture and the other trades, owing to the impact that certain changes to these assumptions may entail.

Most artefacts are preserved under good conditions for relative humidity values between 30% and 60%, provided that the variation between these two extremes is gradual, i.e., greater than a few weeks. It should be noted that these values are perfectly consistent with human thermal comfort.

On the other hand, a very strict control of ambient conditions, particularly relative humidity, entails very high costs, both in terms of the initial investment and, especially, future operational costs.

Therefore, we propose some compromises in order to optimise the initial investment/future operational cost ratio, without neglecting the aims defined for the HVAC design: to simultaneously preserve the collection, thermal comfort and indoor air quality.

The systems are not over-sized and give priority to their centralisation, with an appropriate scaling of the thermal power ratings required, either for cold or heat, and a careful selection that favours energy efficiency and safety in the environment.

The functional organisation of the buildings is duly taken into consideration, allowing for an appropriate hierarchy of ambient treatment levels, including from a perspective of ease of operation and maintenance.

## Energy Systems

The energy systems were envisaged according to temperature and humidity requirements, the balance shown as regards thermal heating and cooling loads, but also the specific use of the spaces.

As a basic solution, we propose the use of refrigerated and heated water production groups, commercially called heat pumps by air condensation of the air/water type.

In view of the different usages, both in terms of time and function, and also the possibility of operation by third parties, eight heat pumps are envisaged, albeit with totally different power ratings and independent operation. The most powerful ones are linked to the ambient treatment of the Exhibition Hall, with the exception of the cafeteria and the storage area, which have a different, dedicated small-scale heat pump. The other four heat pumps are also small-scale and are associated with the auditorium, the administrative offices, the restaurant and the store in the Annex, respectively.

In the case of the two heat pumps for the Exhibition Hall, bearing in mind that this is a building with large spaces and rather demanding humidity control requirements, there are heating needs throughout the year, albeit residual in the colder season, so we considered it appropriate to use energy recovery by thermal heat rejection from the cooling of the heat pump condenser in the chiller version.

Conversely, due to the same need for humidity control and any cooling requirements, even in winter, when the heat pump is working in heating mode, the cold thermal rejection energy from the heating of the evaporator is also recovered, thus increasing the global efficiency of the energy system.

The heat pumps, interconnected to the various items of equipment, promote environmental cooling and heating and air de-humidification, as well as air re-heating in the cooling period.

Solar energy is the basis for the preparation of sanitary hot water for consumption through the use of thermal solar panels, in compliance with national legislation, RSECE, but also the latest European Directive on targets to be attained for renewable energies.

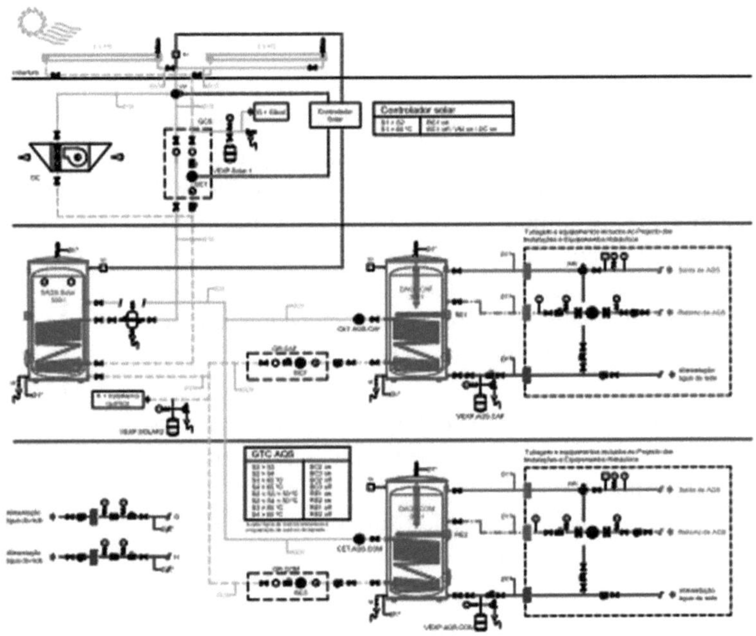


**Fig. 86** Heating system scheme

## Environmental Treatment Systems

The environmental treatment systems are designed in an integrated manner, seeking energy efficiency and sustainability and minimising their impact on the environment, in order to promote:

- the temperature and humidity conditions and the level of filtering defined, both for thermal human comfort and for the preservation of the collection;
- the indoor ambient air quality, i.e., by ensuring efficient ventilation.

The main thing is to simplify solutions and to optimise the technical and economic relationship that will impact both the reduction of the initial investment and especially the decrease in energy consumption and future operational costs.

Without elaborating too much on the systems envisaged, below, we summarise the solutions proposed for the most relevant spaces with regard to the functions of the building:

- Exhibition spaces – In view of the geometry of the spaces and, particularly, the considerable ceiling height, the environmental treatment is jointly ensured by radiant floors and specific air treatment units, both with operation and temperatures very close to set-point temperatures, decreasing the thermal stress on the exhibits.

This combination, associated with air blowing at a relatively low level, enables a volume control to be created that guarantees comfort conditions for all of the spaces;

- Auditorium – Environmental treatment is ensured by a displacement-type solution, taking advantage of load transfer by natural convection to unoccupied higher levels.

Thermally treated air-blowing is done under the boilers at a very low speed and with a low temperature gradient, ensuring indoor air quality and thermal load removal, with extraction being carried out at a high level;

- Administrative area and shop – Thermally treated new air-blowing is utilised, ensuring indoor air quality and local terminal units for removal of the thermal load;
- Restaurant and cafeteria areas – Heat pumps associated with new air treatment units are utilised, ensuring the required new air flow for each space for reasons of hygiene and the removal of thermal loads.

Specific ventilation systems are also envisaged, thus ensuring indoor air quality, notably through extraction localised in areas of high pollutant generation, such as the kitchen and the cafeteria;

- Installation Control – For control of the Installations in the New Coach Museum, a centralised technical management system (CTM) is envisaged. The adoption of a CTM system enables systems to be adjusted over time to the actual needs of the building.

This will mean a more efficient use of energy and fewer wasted esources.

**Fig. 87** AVAC machines

**Fig. 88** Floor heating system

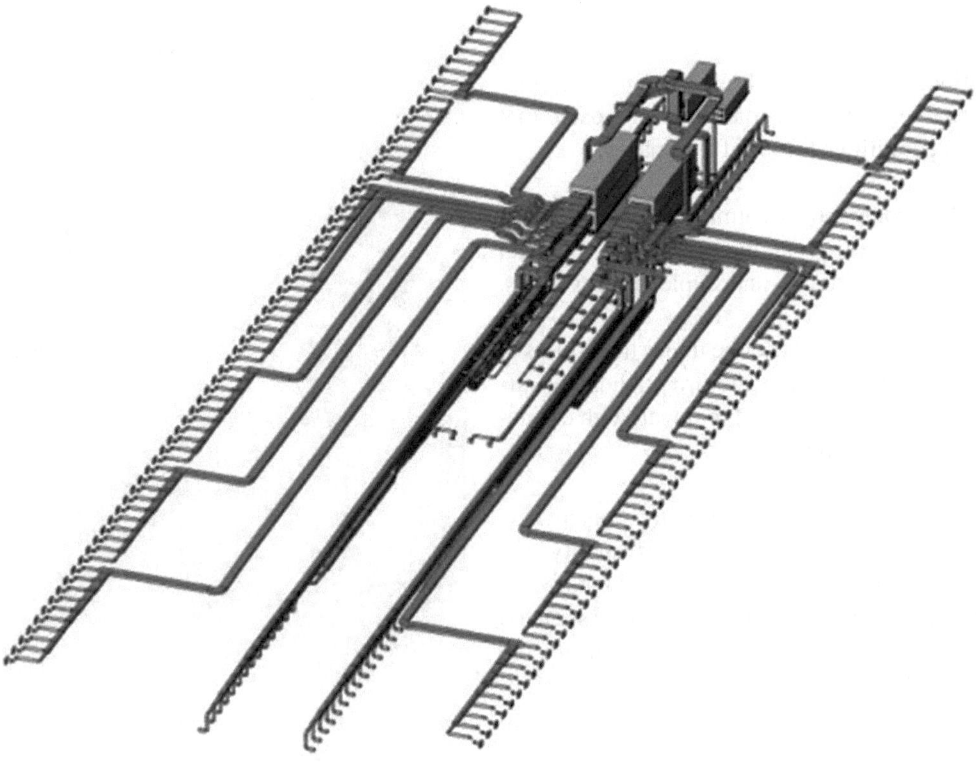

**Fig. 89** AVAC BIM model

**Fig. 90** Ventilation system inside the main wall

**Fig. 91** Main wall with the showcase

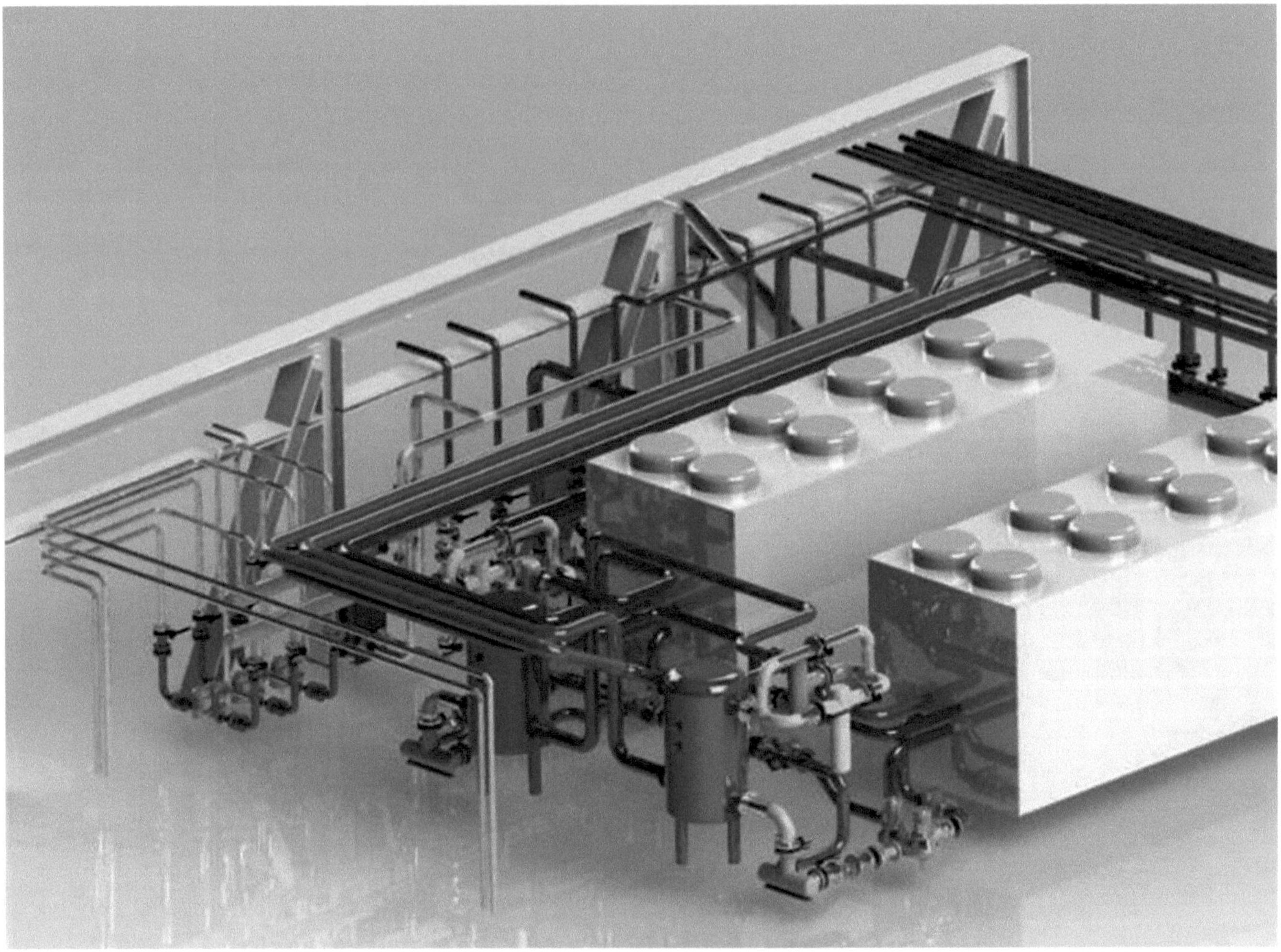

**Fig. 92** AVAC machinery BIM model

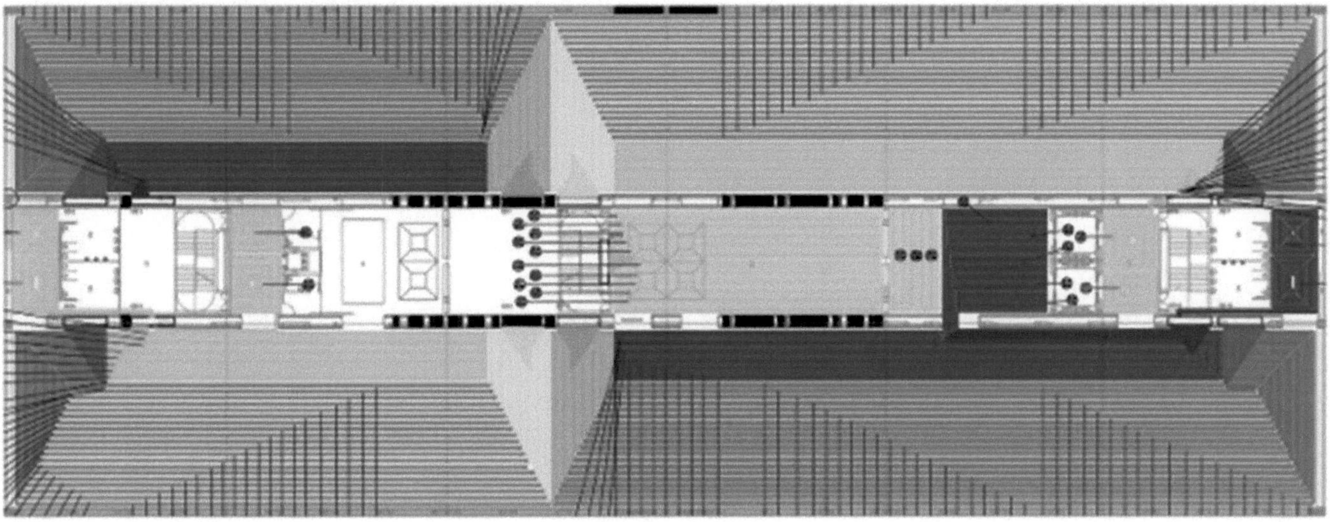

**Fig. 93** AVAC plan BIM model

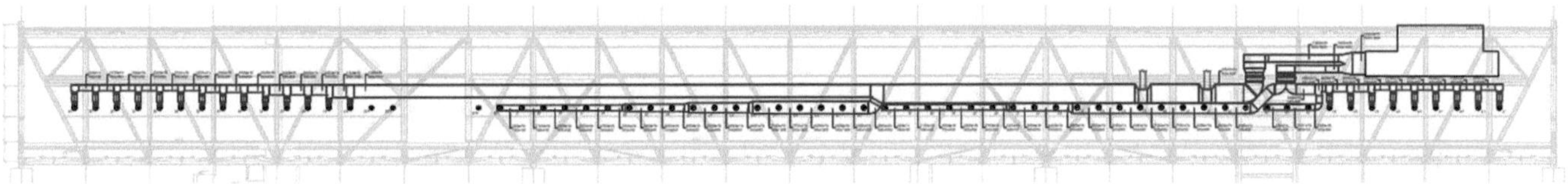

**Fig. 94** AVAC facade BIM model

# Museums Seen Through the Lens of Building Physics

Hugo S. L. C. Hens

The term 'museums' relates to a broad range of buildings with one objective in common: storing, preserving and showing collections of artefacts, ranging from locomotives and cars over furniture to parchments, old books, paintings and other delicate pieces of art.

Before discussing a few aspects related to the building physics and museums, one question must be answered: why me? Why did the co-editor ask me to write a piece on the role of the envelope and the building services involved in the hygrothermal performance of museums? I do, in fact, have a 50 years-long track record in building physics, building services and performance-based building design. My experience with museums, however, mainly concerns heritage buildings, housing artefacts of art. One was Saint Mary's Cathedral in Antwerp, the largest Gothic church in the Low Countries, which displays some precious paintings by Peter Paul Rubens. The cathedral is equipped with an air handling installation, which should keep the indoor climate as constant as possible. This, however, was not quite successful. In the two summer months, the temperature reached 22 °C, while the relative humidity touched 61%. During the three winter months, the temperature dropped to 11.3–14.2 °C for a relative humidity of 64–77%. The daily mean values were also far from 'rather constant'. At the same time, on the rain-exposed side, the grade sits 3 meters above the church's floor level. Through that, the outer wall there was close to capillary saturated, acting as a source of humidity in that way. Another example was the municipal museum in Gouda, the Netherlands. There, the thermal insulation of the heritage building housing the collection, was so bad, mould developed behind the paintings.

Most buildings are constructed to house people. The indoor environment created so focuses on human needs and comfort demands. In museums, things are different. The artefacts that are stored preserved, and shown determine the nature of the indoor environment, not the visitors. One basic concern is preservation. Chemists know that objects of value see their degradation slowed down at lower temperatures. Therefore, the temperature in a museum showing precious artefacts should rather be low. At the same time, relative humidity could be either too high or too low. Paintings, for example, show cracks and wooden pieces of art may spall when the air is too dry. If, instead, it remains too humid, the painted canvas will loosen and stretch[?], while wood swells.

Therefore, when asked to act as building physics consultant, the first question should be: what is the collection that will be stored, preserved, and shown? The answer defines the indoor climate to be provided. Indoor and outdoor climate form the boundary conditions building physicists have to work with. Whereas varying temperature, relative humidity, wind, precipitation, and sun colour the outside climate, the indoor temperature (Temp) and the relative humidity (RH) in a museum should be quite stable. How stable always depends on the collection housed. Also, indoor air quality counts, because pollutants such as sulphur dioxide, nitrogen oxides and many others harm delicate artefacts in the long run, just like solar radiation and the ultraviolet and visible light it contains do.

Looking to temperature and relative humidity, requirements for museums as formulated in the past looked very severe. In the 1970s for example, a constant temperature of 17 °C and a constant relative humidity of 58% were still mandated. Today, a differentiated approach has been adopted, the five levels of indoor climate control proposed in the 2015 ASHRAE handbook of HVAC applications being an example of this.

H. S. L. C. Hens (✉)
KU Leuven Department of Civil Engineering, Leuven, Belgium
e-mail: Hugo.Hens@bwk.kuleuven.be



B. Rangel et al. (eds.), *Museum Technology and Architecture*, Building Research: Design, Construction and Technologies,
https://doi.org/10.1007/978-3-319-76172-5_8

| | Maximum fluctuations and gradients in the controlled spaces | | | |
|---|---|---|---|---|
| Set points, annual average | Class of control | Short fluctuations, space gradients | Seasonal adjustment in set point | Risk of mechanical damage |
| Relative humidity 50%<br>Temperature set between 15 and 20 °C (If loan exhibitions are shown, set points specified in the agreement must be respected) | **AA**<br>Precision control, no seasonal changes | RH ±5%<br>Temp ±2 °C | RH no change, Temp ±5 °C up and down | Nil for most artefacts and paintings |
| | **A**<br>Precision control, some gradients or seasonal changes allowed, not both | RH ±5%<br>Temp ±2 °C | RH up and down 10%, temp up 10 °C and down 5 °C | Small to high vulnerability for artefacts, nil for paintings, photographs and books |
| | | RH ±10%<br>Temp ±2 °C | RH as short, temp up 10 °C and down 5 °C | |
| | **B**<br>Precision control, some gradients plus winter temperature setback allowed | RH ±10%<br>Temp ±5 °C | RH up and down 10%, temp up 10 °C but not above 30 °C | Moderate to high vulnerability for artefacts, tiny for most paintings, most photographs and most books |
| | **C**<br>Prevent all high-risk extremes | RH within 25 and 75% year round, temperature rarely over 30 °C, usually below 25 °C | | High to high vulnerability for artefacts, moderate for most paintings, most photographs and some books |
| | **D**<br>Prevent dampness | RH reliably below 75% | | High plus sudden or cumulative damage for most artefacts and paintings |

Museums showing permanent collections of paintings by famous past masters must go for class AA. Contrastingly, museums that organize temporary exhibitions of actual art may soften the requirements to a B or even C level. Anyhow, the level defines the type of HVAC-system that should be provided and the type of control to install. If, for example, one goes for AA, with the relative humidity fluctuations limited to ±5%, then when large numbers of visitors are expected, the air exchange rate is best kept high. That way, the impact that the perspiration vapour coming off of the many visitors has on the relative humidity in the exhibition rooms, which the dedicated outside air supply air handling units fix, remains low.

The enclosure of any museum, be it housed in an existing monumental building or part of a new construction, has to separate that well-controlled indoor climate from the variability and maladjustment of the outdoor climate. In addition, glazing should be distributed over the facades and roofs in a such a way that no solar radiation touches precious paintings, painted objects, photographs, parchments or books. Artificial lighting must provide the necessary illumination so that colours keep their quality and annoying luminosity contrasts are prevented. This demands a well-studied, carefully designed lighting solution. Additional performance requirements for the museum building as a whole from the point of view of building physics are: energy efficiency, a moisture-tolerant enclosure, excellent sound insulation against outdoor noise.

Of course, the quality of the volumes, of the building's appearance, space organisation integrating storage and preservation demands and exposition requirements with the necessary offices, the restaurant and cafeteria with kitchen, the rest rooms, circulation, and overall functionality also matters, while structural integrity, fire safety, and overall security are equally important. They, however, do not belong to the fields covered by building physics.

Let us now limit the discussion to new museum buildings, (see Fig. 1).

The proper way in which to realize energy efficiency is climate-related. In each climate zone, however, the story always starts with measures at the building's enclosure (also called the envelope) and the fabric level. In heating dominated climates, compactness and a thermally insulated, thermal bridge lacking enclosure are of prime importance. The thermal transmittances to be realized should be baseed on a life cycle cost analysis. Each additional layer of insulation increases the investment cost, while stepwise delivering less energy economy (see Fig. 2). Also positive is good overall thermal inertia. Heavy outer walls, insulated outside (Fig. 3), heavy floors and heavy inner partition walls help to dampen potential indoor temperature fluctuations. In the parts of the building that need windows, these should have well-insulated frames filled with low-e, gas-filled double or triple glazing. Yet, even in heat-dominated climates, rooms with glazed surfaces that face the sun may overheat on sunny days. Minimizing that risk demands the application of effective outdoor solar sunshade systems, which allow for acceptable light transmittance in combination with low solar transmittance. Systems used today include metal screens perforated in such a way that direct sun is blocked but reflected sun enters.

**Fig. 1** New museum buildings

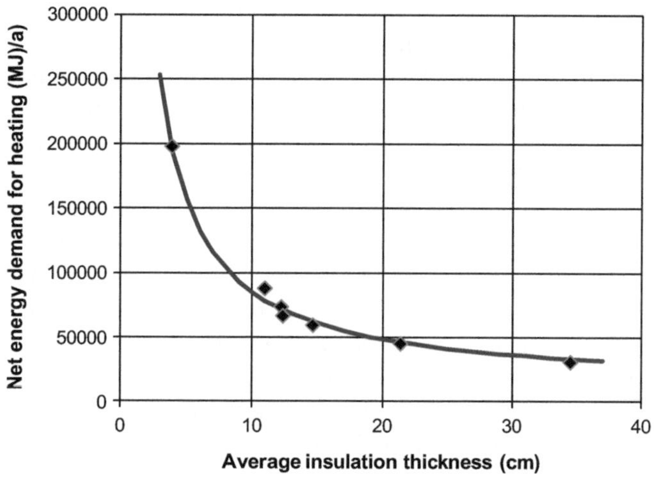


**Fig. 2** Net energy demand of a randomly chosen building, with dependence on the average thickness of the insulation

In cooling dominated climates, thermal insulation of the enclosure is less effective. Some outside insulation should still be provided, mainly because temperature damping of the preferentially heavy outer walls and roofs enclosing the windowless exposition rooms increases that way. Also, a whitewashed outside surface adds benefit, as it reduces solar absorption, resulting in lower outside surface temperatures. Effective solar shading of windows in all rooms that require glazing that sits directly in the path of oncoming sunlight is, of course, of prime importance now.

Besides the building, the HVAC-system chosen also has a direct efficiency impact on the end and primary energy consumed. For the exposition rooms, the choice must go for an all-air, constant volume system. Means to economize include

**Fig. 3** EIFS outside insulation

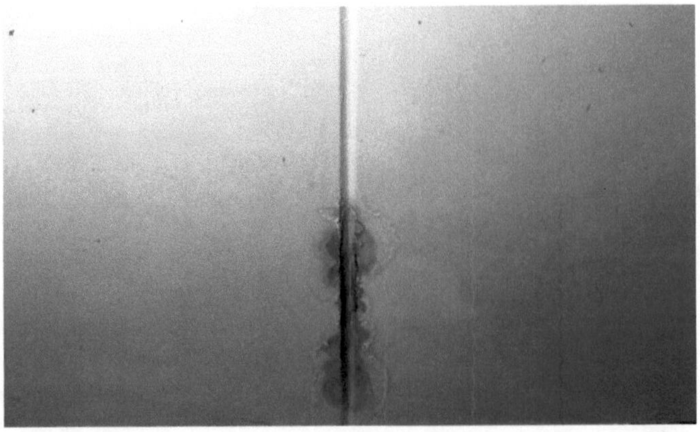

**Fig. 4** Low-sloped roof, improperly designed (inside to outside prefabricated concrete floor elements, glass fibre insulation, wet sloped screed, vapour tight membrane). Summer condensation of building moisture from the screed on the concrete floor elements causes moisture sweeping

heat exchangers between extract and supplied air, the use of heat pumps when possible, looking for highly efficient filters with low hydraulic resistance, steam humidification and careful insulation of ducts and pipes. Dehumidification, is typically done using wet cooling coils at a temperature somewhat below the inside dew point demanded, followed by adequate reheating.

Looking to moisture tolerance, which some call 'the hygrothermal response', the building enclosure faces several sources of liquid water and water vapour. It starts with building moisture. The fabric of every building contains surplus water at the start. Drying must be possible without creating any harm, such as that shown in Fig. 4. The basic rules to be respected are: never switch wet layers between vapour retarding layers, continue heating and ventilating the building outside of visiting hours.

A second problem is rising damp, a source involving ground water, as well as sinking rainwater. The problem is easily solved by the inclusion of a waterproof layer above the grade. Basements, which often serve as storage rooms, should be kept dry, if necessary, by applying an inside or outside waterproof encasement. In many climates, precipitation and wind-driven rain are the main sources of liquid water. Avoiding seepage and rain penetration is done by providing overhangs when aesthetically preferred, but mainly by designing and building the enclosure as a one- or a two-step rain control. One-step means the outside finish has to provide rain-tightness through its ability to act as a drainage plane. Low-sloped roofs and EIFS-insulated massive outer walls are examples of this. A two-step solution splits the outer walls into a veneer or outside cladding and an inner leaf with a non-capillary layer in between. The veneer or cladding acts as a rain buffer and outside drainage plane. The non-capillary layer in between, which could as a well be an air cavity as an insulating fill, allows the backside of the veneer or cladding to act as a second drainage plane without transferring water to the inner leaf. The inner leaf ultimately provides air tightness, wind tightness, structural integrity, sometimes thermal insulation, and sound insulation.

The outside and inside air both contain water vapour. During clear nights, its presence outdoors may give rise to condensation on the covering of well-insulated roofs and the outer finish of well-insulated facade walls, including the outside surface of low-e, gas-filled double and triple glazing. While the latter only blurs the view to the outside, the first can induce mould and algae growth on stuccoed surfaces. Excess water vapour indoors, compared to the concentration outdoors, can, in turn, activate mould growth indoors on thermal bridges. Due to the thermal bridging effects caused by the spacers, it may also result in edge surface condensation on low-e gas-filled double-glazing. In addition, when the opaque enclosure assemblies are improperly designed, with air leakages and mismatches in vapour resistance follow-up, the excess could be the reason why very high levels of sorption moisture build up each winter in the layers at the outside or, even worse, why interstitial condensation deposits from in roofs (Fig. 5) and facade walls. Both 'interstitial' phenomena become more likely in cool and moderate climates, the better insulated the building enclosure.

Take an AA-controlled museum environment, where the temperature is 20 °C and relative humidity touches 50%. Water vapour pressure indoors then equals 1169 Pa. In moderate climates, the lowest monthly mean winter temperature noted drops to ≈0 °C, for a vapour pressure of 550 Pa. Long lasting condensation on the outside surfaces will occur when undercooling brings the monthly mean surface temperature down to −1.3 °C. Inside, mould is expected to develop on surfaces that stay colder than 12.7 °C, a value we note on thermal bridges whose local thermal resistance drops below 0.4 $m^2$.K/W. Between brackets, such low inside surface temperatures are also found behind paintings hung on non-insulated masonry outer walls in existing landmark museums. A thermal

**Fig. 5** Insulated pitched roofs, leaky vapour retarder and vapour-tight, non-capillary underlay. The result is an interstitial condensation deposit with run-off and water leakage at the roof edges

resistance of 0.4 $m^2.K/W$, in fact, convenes with the thermal quality that a two-brick thick outer wall offers. Surface condensation inside at the edges of low-e, gas-filled double-glazing, in turn, becomes very likely during any cold day, as the temperature there conforms to follows the instantaneous outside temperature changes.

To conclude, moisture tolerance or, in other words, hygrothermal performance, demands a series of knowledgeable decisions in regard to building enclosure design.

Good sound insulation against outside noise for the exposition rooms is, again, best served by a massive construction with heavy outer walls and roofs and no windows. As visitors are usually quiet when walking through exposition rooms, no special reverberation time requirements prevail and sound-absorbing surfaces are not needed. This, of course, does not hold for exposition rooms where visitors are engaged to interact. In those cases, sufficient sound-absorbing surfaces must be planned, a requirement that often conflicts with a demand for thermal inertia. In rooms that need windows while requiring a very silent indoor environment, one should use low-e, gas-filled double-glazing with excellent noise transmission loss. From the point of view of room acoustics, they demand treatment in accordance with the requirements that are valid for offices, restaurants, pubs, etc. They are typically well-served by a sound-absorbing ceiling. If the museum includes lecture rooms, acoustics that allow the speakers to be intelligible is of prime importance there. This demands a well-balanced combination of a sound-absorbing ceiling and a rear wall with a reflecting surface in front and reflective sidewalls.

To get everything that we have discussed right, good cooperation must grow between the architect and the building physics consultant. Our experience is, nevertheless, that some architects too often overlook these building physics-related aspects and design museums that demand corrective measures afterwards, once built.

# Coordination of Work on the Coach Museum Supported by BIM Methodology

António Ruivo Meireles, Bernardo Salavessa
and Fernando Gonçalves

## Introduction

In October, an imposing building of more than 11,500 $m^2$ was opened in Lisbon, between Avenida da India and Rua da Junqueira, which has the potential to contribute to the transformation of the city: the New Coach Museum. According to the architect, Paulo Mendes da Rocha, the museum will host "the whole extraordinary lyricism associated with coaches, with (...) beautiful (...) Neptunes and angels and gilded figures," and with that in mind, nothing has been left to chance in its design.

The building has a number of idiosyncrasies, most notably the fact that it is a volume of rectangular shapes (see photo below) and broad spaces, and that it is "suspended,"

**Fig. 1** Coach Museum in March 2012. (Photo by Raimundo Constâncio)

A. R. Meireles (✉)
ndBIM Virtual Building, Vila Nova de Famalicão, Portugal
e-mail: armeireles@ndbim.eu

B. Salavessa · F. Gonçalves
MOTA-ENGIL, Engenharia e Construção, S.A., Setúbal, Portugal


B. Rangel et al. (eds.), *Museum Technology and Architecture*, Building Research: Design, Construction and Technologies,
https://doi.org/10.1007/978-3-319-76172-5_9

**Fig. 2** 25 m × 100 m slab finished directly. (Photo by Raimundo Constâncio)

with spans of considerable size required to support art exhibits more than 7 m in length and weighing tons.

There are many stories that could be told about the construction of this building, such as the concreting of a 25 m × 120 m slab that was finished directly. However, with the aim of awakening the construction industry to the potential of the Building Information Model (BIM), the focus of this article will be the contribution of this new methodology to this particular work.

## The Building Information Model

Various forms of simulation have been used throughout history. The wooden models of the fifteenth century in the Renaissance period are an example, as are the diagrams, designs and specifications that have been used for hundreds of years as a means of transmitting information. However, the information contained in these forms is incomplete and fragmented.

The BIM is a new approach that involves the generation and management of a virtual physical and functional representation of a project. The result is a shared knowledge base that will support decision-making throughout the life of the project, allowing you to reduce costs, eliminate waste and improve communication between all of the participants in the value chain. The BIM is therefore the project, but also the simulation of the construction process and its use.

## The Challenge

The suspended body of the New Coach Museum is supported by a dense metal structure with mechanical properties that provide the strength to sustain the tons of various coaches and the thousands of visitors expected to flow through this building, in addition to its own weight resulting from a structural design aimed at providing broad and free spaces.

It is important to stress the challenge of achieving the compatibility of the existing engineering specialties with the placement of hundreds of metres of pipes and tubing, in a confined space in the false ceiling of the metal structure, as shown in the following figure.

**Fig. 3** Metallic structure. (Photos by Raimundo Constâncio)

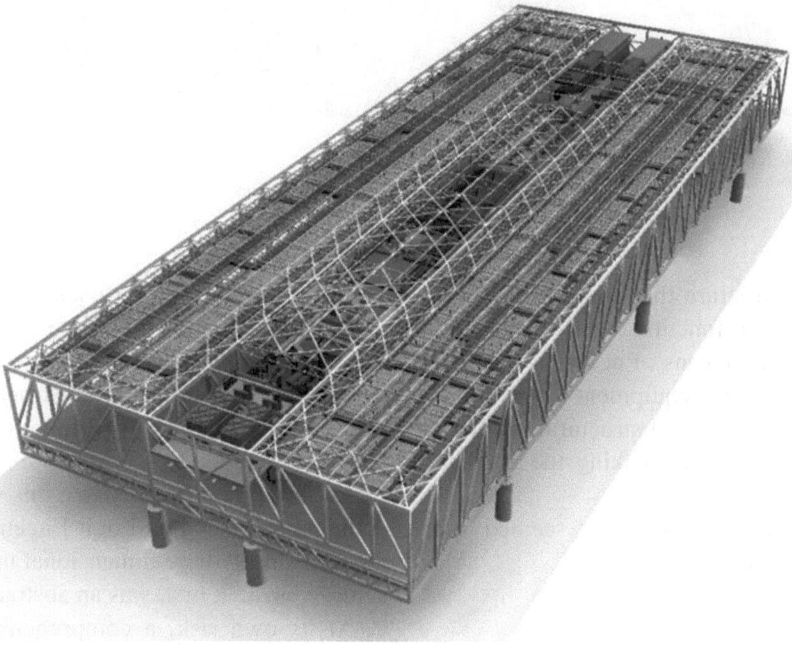

**Fig. 4** Representation of the density of the services networks

Note that although, to begin with, it seemed unthinkable that, in a museum of this magnitude, the various services could remain on view, in practice, the concept of the architect, Paulo Mendes da Rocha, was that they would be visible through a false ceiling constituted by a non-opaque metal grating. This will enable the public, in general, to see the complex tangle of ducts, conduits and cable trays that facilitate different services, increasing the need for special care in the layouts and in their presentation and finishes.

Between this ceiling and the sheet metal roof, through the lattice of the metal structure, there are the HVAC installations (pipes and tubing- better conducts instead of tubing), cable trays for electrical, communications and safety power cables, luminaires, switchboards, the fire-fighting and sprinkler network, the drainage system for rainwater from the roof, and many others.

The metal grid encloses the space where all of these infrastructures are found and it comes close to touching the

**Fig. 5** Physical space for the service installations. (Photo by Raimundo Constâncio)

lower level of the lattice structure that supports the roof. Simultaneously, it has a similar function to the "gantries" in theatres, and can be used as a means of access for maintenance of the infrastructure and the equipment that sits on it, which explains the need for increased structural reinforcement and four hangers per section, making the available space even more difficult and restricted.

In other words, the space that, at the start, seemed to be ample turned out to be cramped.

## The Solution – Introduction of the BIM to Assist in the Coordination of the Project

In this sense, a fruitful collaboration between the engineering team and the builder became paramount. It combined design knowledge with practical know-how, with both sides working together to further optimise the space and mitigate risks.

Early on, the builder, through the detailed analysis of the drawings, realised the risks of the various specialties not having been properly brought into line with each other and that it would be necessary to invest in their integration.

The designer himself, conscious of the limitations of the available space, included some cross-sections in his project that illustrated the layouts for the services and their interconnection with the metallic structure, but this was unfortunately not enough to allow for the resolution of all conflicts.

Aware of the difficulty that this would entail in regard to communication, Mota-Engil Engenharia decided to use the most advanced three-dimensional modelling technologies to achieve what, at first, was an abstract analysis.

At its own risk, a comprehensive study was initiated with the BIM of three-dimensional modelling of all engineering specialties over six consecutive weeks, in order to identify collisions and to propose resolutions for them to the designers.

Beyond the clear advantage that the model gave by helping with the analysis and clarified perception of the project, about seventy errors and incompatibilities were identified throughout the work (see the extract from the document below) that were presented to the designers for resolution, triggering the process of discussion and adaptation of the designs.

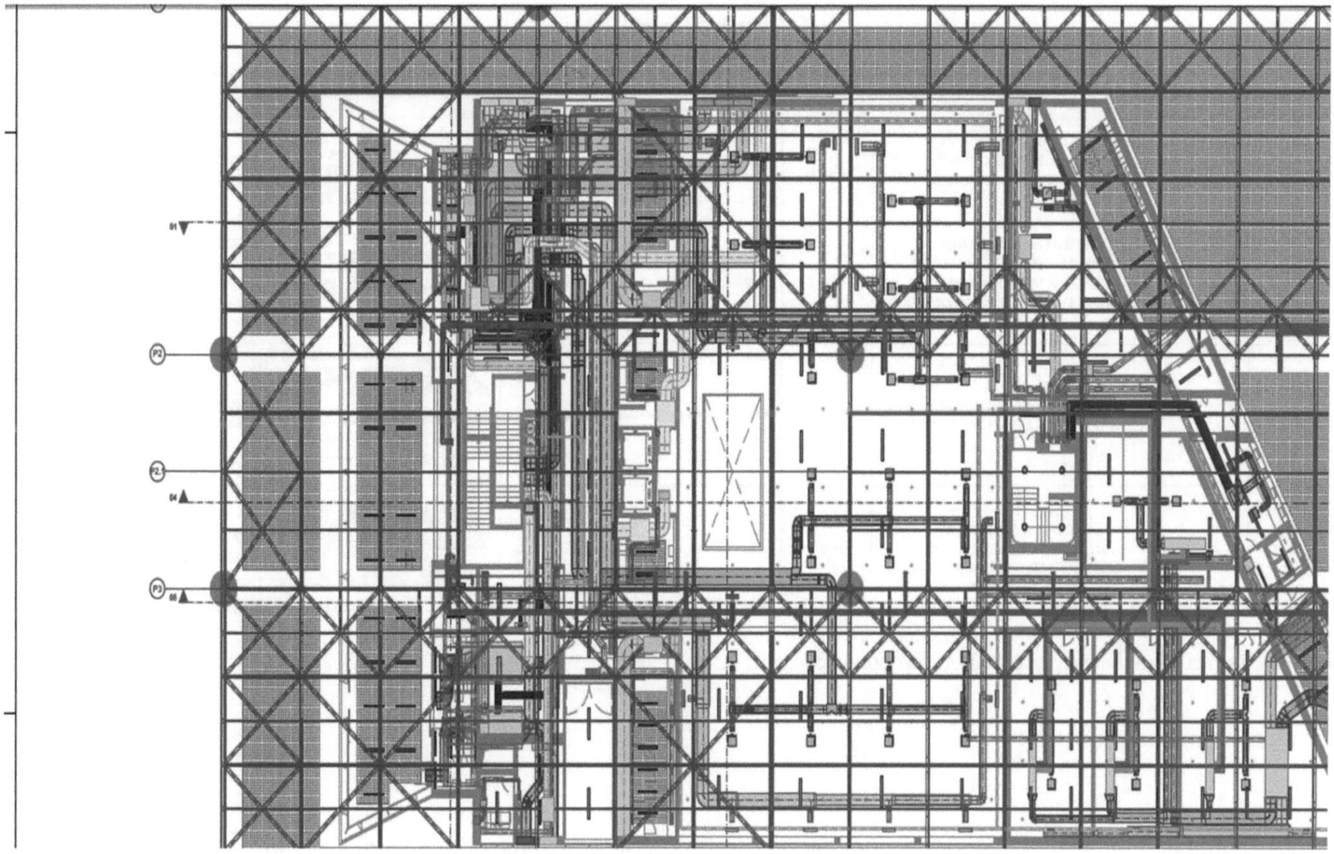

**Fig. 6** Complexity of the networks in the drawing supplied by the designer

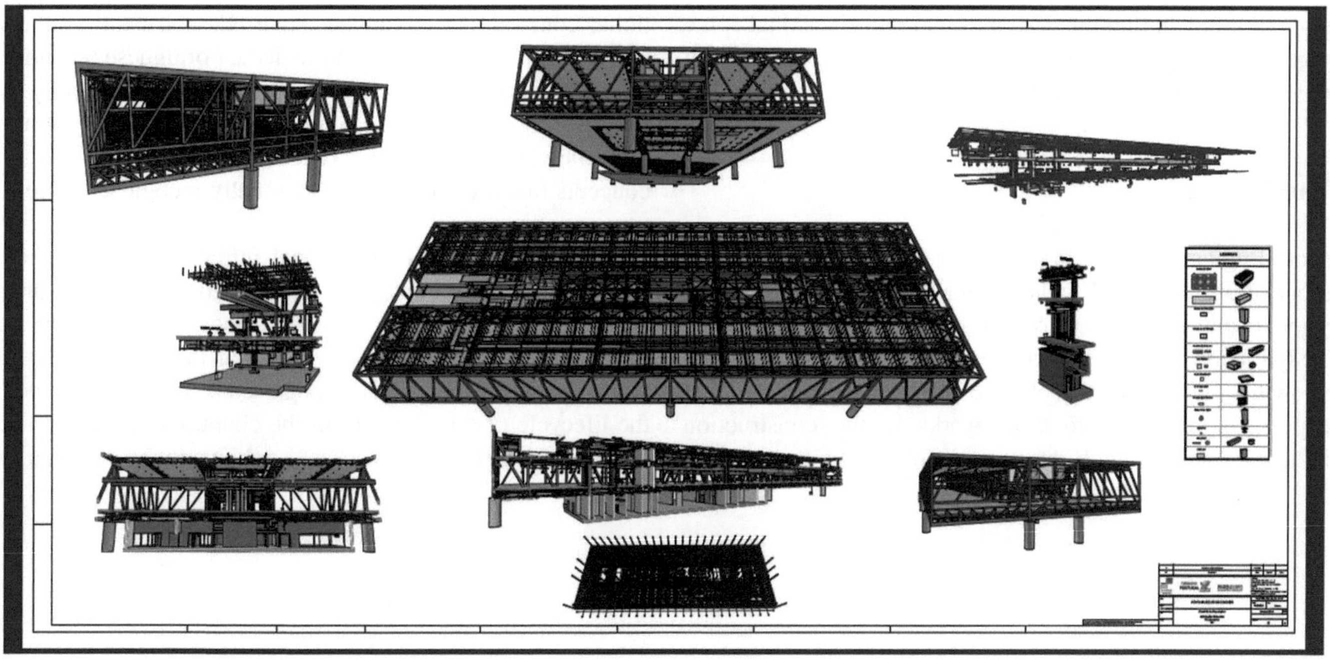

**Fig. 7** Example of drawing presented to the designers

| CC12209 - NOVO MUSEU DOS COCHES<br>Relatorio de Incompatibilidades | | | | | | | |
|---|---|---|---|---|---|---|---|
| ID | Especialidades | Descrição | Piso | Eixos | Desenhos | Screenshots | Decisão de Modelação |
| P1_001 | EST/AVAC | COLISÃO ENTRE A CONDUTA DE DESENFUMAGEM E CONDUTA DE EXTRACÇÃO E COM A ESTRUTURA METÁLICA | Piso 1 | PB2;PB3 / P1;P2 | FT0201MECPEPL101A |  | MODELOU-SE CONFORME O TRAÇADO INDICADO NO PROJECTO DE AVAC. |
| P1_002 | EST/AVAC | COLISÃO ENTRE A CONDUTA DE DESENFUMAGEM PRESSURIZADA E COM A ESTRUTURA METÁLICA. PROJECTO INCOERENTE COM O DIMENSIONAMENTO INDICADO EM PLANTA. | Piso 1 | PB2;PB3 / P1;P2 | FT0201MECPEPL101A |  | MODELOU-SE CONFORME O TRAÇADO INDICADO NO PROJECTO DE AVAC. mODELOU-SE CONFORME O DIMENSIONAMENTO INDICADO EM PLANTA |
| P1_003 | EST/AVAC | COLISÃO ENTRE A CONDUTA DE EXTRACÇÃO E A ESTRUTURA METALICA | Piso 1 | PD1;PD2 / P1;P2 | FT0201MECPEPL101A |  | MODELOU-SE CONFORME O TRAÇADO INDICADO NO PROJECTO DE AVAC. |
| P1_004 | EST/AVAC | COLISÃO ENTRE A CONDUTA DE INSUFLAÇÃO COM A CONDUTA DE RETORNO E A ESTRUTURA METÁLICA | Piso 1 | PB3;PB4 / P1;P2 | FT0201MECPEPL101A | 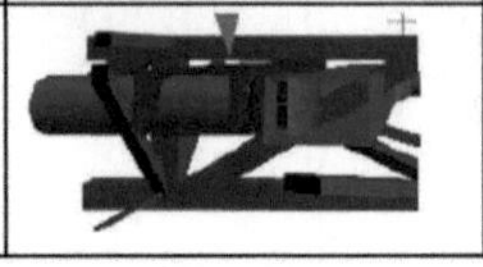 | MODELOU-SE CONFORME O TRAÇADO INDICADO NO PROJECTO DE AVAC. |

**Fig. 8** Extract from the incompatibility report produced

The collaboration between everyone was very fruitful and, as a result, improvements were made to the design and preparation of the layouts. This consequently led to fewer site problems, and therefore lower extra costs. We admit that we could have gone further if everyone involved in this collaboration, including the designers and ourselves, had come together to take greater advantage of the BIM modelling tool in resolving these inconsistencies.

## The Future

Mota-Engil argues that, in future projects, the process of harmonisation with BIM tools should be achieved in a stage prior to that of construction, mitigating the burden on the design team during the construction support phase and allowing the builder to optimise planning. The traditional expression, common to most works in the construction industry, "on site, with the pipes in place, we'll see" runs contrary to the principle that we aim to follow, i.e., to anticipate and resolve problems at an early stage of the project, decreasing the cost to the customer.

Mota-Engil has invested in exploring the concept of BIM over the past three years, and, together with other international construction companies (like Hartela, Peab, Webcor and Hathaway Dinwiddie) and with the Portuguese Scientific and Technological System (University of Porto Faculty of Engineering, Higher Education Technical School, Minho University, among others), has developed methodologies and concepts that are already internationally recognised as relevant and pioneering.

In November 2011, a national initiative was launched, called BIMFórum Portugal, supported by companies on a voluntary basis, which aims to boost the concept in Portuguese-speaking countries and spur the development of working standards, which is recognised by all stakeholders in the lifecycle of a project (from the client, the designers and constructors through to the user), as shown in the next image.

**Fig. 9** Scheme ilustrating the importance of BIM in the building lifecycle

The potential of BIM was not fully realised in the presented case of the New Coach Museum. While it did pinpoint the project's incompatibilities, which is the most visible face of BIM's uses, this is probably not its most advantageous trait. We could list the numerous advantages to each stakeholder, but in general terms, we can say that there are dimensions that benefit all involved. The 3D dimension refers to COORDINATION/QUALITY, in other words, using the model to reconcile the various specialties and support consistency between drawings and written documents. The 4D concerns the TIME factor, i.e., the ability to plan the project based on the simulation of various scenarios and risk analyses. Finally the 5D involves the COST factor, the ability to make informed decisions, i.e., to estimate a project more accurately by mitigating the possibility of erroneously quantifying certain tasks.

# Hygrothermal Behaviour Control in the Crawlspaces of Historical Museums

João M. P. Q. Delgado and Ana Sofia Guimarães

## Introduction

Building conservators and curators are now willing to accept the importance of controlling wider temperature and relative humidity fluctuations. The key is to control the drift of relative humidity and allow temperature to fluctuate within a wider span. While each situation must be assessed, temperatures ranging from 7 °C in winter to 27 °C in summer, with relative humidity from 75% in winter to 30% in summer, are possible. Seasonal shifts and drifts must be controlled to avoid sharp peaks or drops. Most collections can work within these parameters or be placed in climate controlled cases or storage. Buildings generally do well within these parameters, but buildings prone to rising damp may show damage below roughly 10 °C. Dimensional changes in building materials, particularly assemblies made up of different materials, can occur above 27 °C or if the building is allowed to freeze. Paintings, wood panelling and other fragile materials can be severely affected in winter if the humidity levels drop too low and the heating is too high. Likewise, mould may appear on leather and wooden surfaces if the humidity gets too high, so monitoring and controls are necessary in many situations. The dew point cannot be ignored, because condensation is a major concern. Fan ventilation systems controlled by humidistats can provide low-tech mitigation for buildings that experience high temperatures and high relative humidity.

Historical Museums usually possess crawlspaces, as this was a traditional tool of construction in the past. Those spaces can be correctly used to help maintain favourable internal climatic conditions favourable.

In recent years, several crawlspaces have been treated in Portugal for mould and moisture problems. Crawlspaces have been around for a long time all over the world. In the past, the first floors of wood were installed with an air layer between the soil and floor. Some of the most important advantages of crawlspaces are: the cost effectiveness, the easy access to water and heat installations (useful for low cost repairs), the fact that the floor is secured from capillary suction from the ground and that the evaporated water can be removed by ventilation.

However, crawlspaces are one of the foremost causes of moisture and mould problems in houses, which can be a serious health issue for the occupants (causing allergies and sicknesses) and result in costly renovation [6, 9, 16, 17, 20, 26]. Over recent years, indoor air problems related to mould growth and moisture problems in crawlspaces have typically been considered for both warm and cold crawlspaces.

In the beginning of the 1950s, crawlspace ventilation became mandated by building codes. The basis for the code requirements was the combined idea that the primary source of crawlspace moisture was evaporation from the floor and that ventilation would allow for dissipation of this evaporated moisture.

To understand the moisture problems of ventilated crawlspaces, the heat and moisture transfer processes must be understood. The evaporation of ground moisture depends on the mass transfer over the ground surface and the moisture transfer in the ground, and the air change rate is mostly driven by wind [24].

During most of the year, moisture is evaporated from the ground surface, but in the months when the outdoor air temperature is higher than the air temperature of the crawlspace, the moisture transport may be reversed, and in some cases, this phenomenon can cause condensation. A great number of researchers [24, 25, 27] have stated that the main reason for the moisture is the uncontrolled ground moisture evaporation and a lack of air change. However, for cold crawlspaces, the process of drying by ventilation is problematic. The effect of ventilation can be both drying and wetting, depending on the temperature and moisture content in the outdoor air compared to the crawlspace air.

J. M. P. Q. Delgado · A. S. Guimarães (✉)
CONSTRUCT-LFC, Department of Civil Engineering, Faculty of Engineering, Porto University, Porto, Portugal
e-mail: jdelgado@fe.up.pt; anasofia@fe.up.pt





B. Rangel et al. (eds.), *Museum Technology and Architecture*, Building Research: Design, Construction and Technologies,
https://doi.org/10.1007/978-3-319-76172-5_10

In conclusion, crawlspaces are a cheap and traditional manner of building a house's foundation, however, if they are not properly constructed, the reparation costs can easily exceed the money that was initially saved [4, 5]. The key issue is to prevent water from entering the crawlspace, but the ventilation is always necessary if any moisture evaporation occurs.

Mould growth has been found on many different types of material. The most common ones are organic materials such as wood and materials with organic components such as gypsum boards with paper surfaces. The mould growth on wooden material has been the subject of experimental research for a long time, with the aim of describing the material's response and finding the critical conditions for mould growth on surfaces of different materials [2, 18, 29].

The wood rotting process is divided into two phases [28]. In the first phase, called the "initial stage," fungi in the air stick to the wooden surface and their hypha grow into the pores in the wood. When the fungi have grown to a certain size, they begin to "eat" the wood, so to speak. This is the beginning of the second phase, entitled the "growth stage" [22, 23].

The most important factors causing mould growth in a crawlspace are relative humidity, temperature, nutrition and pH. In practice, relative humidity is the most significant factor that causes mould growth in crawlspaces. The limit value for relative humidity in crawlspaces is usually considered to be from 75% to 80% [30]. Relative humidity and temperature are strongly linked to one another. The thermal mass of the structures and the ground soil affects the temperature behaviour of the crawlspace. Unfortunately, it is quite common to get levels of relative humidity near the saturation. It is important to keep in mind the psychrometric diagram in avoiding occurrences of condensation in the spaces and, consequently, pathologies.

## Rising Damp in Museums

The problem of damage to historic masonry due to salt crystallization by rising damp is well-known from sites such as Venice, Italy and the Tower of London [1, 3]. Rising damp is one of the leading causes of building degradation, in particular, in older constructions like monuments and museums. This occurs only on the lower floors, where the walls and foundations are in direct contact with water or damp soil. In older buildings, the situation is recurrent, motivated either by a lack of a prevention system or by the characteristics of the building elements. The presence of this phenomenon in buildings can lead to accelerated deterioration of building elements in contact with the ground, and may even promote unhealthy conditions for users [12].

Although rising damp manifests itself frequently, it is a complex phenomenon, exhibiting unpredictable behaviour. The treatment solutions commonly used often do not present as sufficiently effective, especially in buildings with very thick heterogeneous walls. Choosing the incorrect intervention solution may result in the inefficacy of treatment and, under some circumstances, the deterioration of building conditions. This can be related to the difficulty of consulting information associated with corrective measures for this phenomenon, largely because of the significant dispersion of existing information on the topic.

This problem is found not only at the technical level and not only in regard to issues such as features, application methods and advantages and limitations, but also in the selection process of the best intervention techniques from all of the existing options for a potential user. In this way, it is clearly important to know the set of possible solutions in detail, including their respective advantages and limitations [11].

The systematization and disclosure of information about the techniques for treating rising damp in the form of files/catalogue gives professionals, technicians and students instant access to the most relevant information on this matter. In each case, the most appropriate treatment solution can be selected, with a predictable outcome.

## Treatment Techniques

The interventions to be carried out in the treatment of rising damp should be properly planned before their implementation. Definition of the factors that may affect the success of the application is crucial; these include: material constitution, characteristics and thickness of affected building elements, the initial water content, the interior conditions of the building and its surroundings, as well as the respective location. With planning and research conducted, the criteria are established for defining the proper technique in each instance, thus obtaining greater success in solving the problem. Due to the difficulty and variability of factors, some unpredictability exists in the results. The treatment selection should be supported by a complete diagnostic of the origin of the problem, by studying expected effectiveness and correlating that with the cost of the intervention.

The usual treatment techniques for rising damp are based on four principles of intervention: Stop the access of water to the walls; Remove the excess of water in the walls; Stop the ascension of water in the walls; and Conceal anomalies.

In Table 1, the different techniques applied for the treatment of rising damp in building walls, in particular, in walls of historical buildings used as museums, are summarized.

**Table 1** Summary of the treatment techniques for mitigating rising damp

| Treatment technique | Description |
|---|---|
| Stop the access of water to the walls | In order to stop the access of water to the walls, peripheral drainage may be a possibility. These systems do not allow the present ground water to contact the wall, thereby representing a principle that does not act directly on the phenomenon. It will not interfere in the decrease of the height reached by the wet front wall when present. The element will not be supplied by the source, which should help to reduce the problem. These systems tend to be used in cases of damp prevention, however, their implementation in the rehabilitation of buildings is achievable when you want to waterproof the existing wall and there are no technical or financial restrictions. |
| Remove the excess water in the walls | The systems related to this principle act directly on the element, by decreasing the presence of rising damp in the element. However, they do not treat the source of the damp, resulting in the constant presence of the phenomenon of capillary rise. |
| Atmospheric drainage siphons | The principle behind atmospheric drainage siphons is the fact that damp air is heavier than dry air. Knappen believed that inserting oblique drainage tubes into walls would release damp air (coming from inside the wall), thereby facilitating the wall-drying process. |
| Electrical osmotic | This is an old technique that delays the rising of the water by creating an electric potential against the capillary potential. This technique is no longer popular, because it is not considered to be effective. |
| Electronic impulse | This is a completely electronically-controlled procedure. The system generates fine impulses, which change the di-polarity of the water. This is due to the fact that the negatively loaded earth now draws the water. The masonry dries up because the humidity is now prevented from rising. |
| Stop the ascension of water in the walls | The principle of stopping the ascension of water in the walls is possibly the most effective system for the treatment solutions of rising damp (when properly performed). This principle, just like the previous techniques, is not at the origin of the capillary rise phenomena. |
| Reducing the absorbent section | This technique consists of diminishing the absorbent area by replacing part of the porous material (wall) with air pockets, thereby not only reducing the amount of water absorbed, but also increasing evaporation. |
| Physical barrier | A watertight material (bitumen, polymer-based mortar, corrugated sheets of stainless steel or lead sheets) is inserted into the wall's buried section to prevent water from migrating to the upper levels. |
| Chemical barrier | Whereas the above-mentioned techniques of creating a damp-proof cut-off consist of a physical barrier, this technique introduces a chemical barrier. The chemicals can be introduced into the walls by diffusion, injection or applicator gun through holes drilled at intervals to ensure that the chemical barrier covers the entire width and length of the affected walls. |
| Ventilating the wall base | Creating ventilated peripheral channels, in addition to diminishing water contact with porous walls, increases the evaporation of absorbed water. This evaporation takes place below the ground level. Installing a hygro-regulated mechanical ventilation device can increase this system's effectiveness. |
| Concealing anomalies | On the principle of concealing anomalies, the associated systems do not act on the origin or prevent rising damp in the wall, but rather hide the manifestation and also increase the evaporation surface of the various elements. |
| New wall separated from the original wall with a ventilation space | When the causes of rising damp cannot be eliminated, we can decide to put up a new wall separated from the original wall by a ventilation space, which is a type of damp-proof course system concealing the anomalies. |
| Coating with controlled porosity | Applying outer coatings that promote the evaporation of humidity from inside the walls and that impede salt from crystallising on the outside is a technique that conceals the problem. |

## HUMITECNIC – Catalogue

The creation of this catalogue (HUMITECNIC) was done with the intention of collecting and structuring information about the methodologies and treatment of rising damp in buildings by aggregating all of the existing information in the form of patents, treatment systems, products and materials. HUMITECNIC intends to describe the files/catalogue in an organized and accessible manner. Given the multiplicity of offerings in the construction market, by having all of the techniques associated with rising damp treatment, the difficulty in selecting and scaling the intervention technique most appropriate in each case is reduced.

The virtual page HUMITECNIC thus presents the following link, which allows access to the website, "*rolandojgfreitas.wix.com/Humitecnic*". The only files/catalogue HUMITECNIC focuses on are those regarding the treatment of constructive elements with existing pathologies associated with the phenomenon of rising damp, describing the different techniques and known treatments. Although prevention is presented as the best way to combat rising damp, it is important to know techniques of treatment for cases when your system prevention is ineffective or absent, as often happens in old buildings.

Taking into account the current state of the construction market in Portugal, it is clear that we are witnessing a dedicated investment in the rehabilitation of buildings. Having this option, construction gained a source of information in a market that appears to have an obvious need to learn rehabilitation techniques and methodologies for treatment of the phenomenon of rising damp, typically present in older buildings.

These files allow the user to know the different treatment methods typically used in the construction market. HUMITECNIC discloses the active principles that describe the technology used in the installation process, the most relevant properties, and the applicability and limitations of the different techniques. These claim to be a support for any type of user, within the sector or not, who wishes to know different solutions to this problem.

The knowledge in the catalogue allows the user to evaluate the characteristics and limitations of the assorted treatment techniques, so as to select the most appropriate. The catalogue supports the user by not limiting its information to that of the more popular, typically recommended interventions, thereby preventing an all too common lack of information and knowledge of other techniques and conditions. The main goal was to provide a tool that contributes to adequate intervention in the treatment of pathologies associated with such a phenomenon, not only serving users at the academic and technician levels, but also any user requiring information about the existing treatments. The user is therefore not limited to the treatment techniques that he/she is knowledgeable about, being able to compare solutions and evaluate applicability.

The HUMITECNIC catalogue consists of a set of 36 files (an example is shown in Fig. 1) prepared based on research into the constitution of the treatment solutions, and whether or not the products and/or materials are patented or not patented. Catalogues of various national and international companies were analysed. The companies work with different techniques for treating rising damp, which haved subsequently been adapted into the HUMITECNIC catalogue format.

In some cases, there are systems or solutions found in the respective record that were based on existing literature. The catalogue does not specify treatment techniques for trademarked products. It is intended to produce a document in a simple and synthetic way that is interesting and useful. These are presented in a systematic format to makes them easier to read, so the user can quickly see the different fields of interest. The fields that characterize the catalogue are necessary so as to inform the user of their choices of different types of intervention, product and system. Fields like description and characteristics, installation process, representation, properties, application and limitations were selected to best characterize the treatment.

The files of the treatment techniques for rising damp are grouped according to their intervention principle: prevent access of water to the walls, remove the excess water in the walls, stop the ascension of water in the walls and conceal anomalies. Knowledge of the intervention principle is crucial for choosing the most adequate solution, ensuring greater effectiveness.

The subdivision of treatments by principles is planned, but it is not intended to individualize the different methodologies. But before a connection among different interventions can be shown, depending on the principle, it may need to be ensured that they represent greater effectiveness in treating rising damp together. This is because, when combined, they can remove the problem completely.

Dividing up the four active principles allows the user, after making a correct diagnosis, to consult the catalogue in a more organized and simple manner, understanding the order of each treatment. The user does not need to examine all treatment techniques to realize the most adequate one for treating the problem or to avoid using inappropriate treatments to resolve the same situation. Assigning a colour for easier identification of their type is the easiest way to identify these principles.

In the treatment of rising damp, it is possible to use strategies that are mainly based on two approaches: Products that minimize or eliminate the phenomenon and/or systems that tend to alter the conditions of the physical system *potentiate*. A product is characterized by a substance or group of substances manufactured through chemical or biological reactions that are applied directly in the treatment of the affected element. Treatment systems are characterized by either techniques or actions, along with machinery and equipment, which all come together to allow for the treatment of rising damp.

## Crawlspaces of Historical Museums – The Case Study of the Egas Moniz Museum House

In this work, a case study of the Egas Moniz museum house (see Fig. 2), which experienced wood rot problems within its the crawlspace, is presented. The Egas Moniz museum house is in Avanca, Estarreja, about 50 km southwest of Porto. The building dates back to the eighteenth century, but was completely rebuilt by Egas Moniz (the first Portuguese Nobel Prize winner) in 1915 with the help of the project's architect Ernesto Korrodi.

The building has been requalified and named the Egas Moniz museum house. The building was opened on July 14, 1968, and classified as a Building of Public Interest in 1997. The house has architectural interest and also contains Professor Moniz's valuable art collection of furniture, porcelain, glassware, Portuguese paintings, engravings, sculpture, gold items and tapestries. The house, envisioned[?] by Korrody in 1915, along with the neighbouring Quinta do Marinheiro, was converted into a museum house, according to the will written by Egas Moniz (Portuguese Nobel Prize in Medicine 1949).

**HUMITECNIC – STOP THE ASCENSION OF WATER IN THE WALLS**

**Chemical barrier – Diffusion – Liquid synthetic resin**

Ref.:BQ.D-1

**Description and characteristics**

- Very thin liquid synthetic resin;
- Penetrates deeply into even the smallest capillaries and pores in building materials;
- Low density;
- Surface tension lower than that of water, displaces the water in the capillaries;
- The curing of the injected product is independent of the drying of the masonry;
- After is full cure, remains flexible, does not decay or decompose, acts neutrally, and does not effloresce;
- Resistant to all of the usual corrosives in masonry, such as acids, alkalis, and salts;
- Does not affect steel reinforcement;
- Water repellent;
- Can also be applied using low pressure injection systems;
- In cases of high degrees of moisture penetration or in the case of very dense building materials, it is possible that a longer exposure time can be necessary.

**Installation process**

1. Remove the old coatings to treat the walls about 50 cm above the moisture mark;
2. Repair any cracks or cavities for better distribution of the product;
3. Drill horizontal holes in the lowest horizontal joint with a depth of 5 cm less than thickness of the masonry;
4. Clean the holes by flushing with compressed air or briefly with water;
5. Measure and cut the capillary rods, at least 7 cm longer than the depth of the drill hole;
6. Fill the supply tank with water twice in short intervals in order to achieve a rapid swelling of the capillary rods;
7. Place the liquid synthetic resin, after about 15 minutes, in the suction angle's clamping device, so that the supply tank fills up with product;
8. Remove the cartridge after approximately 24-48 hours;
9. After applying the injection, the capillary rods can remain in the masonry. Protruding ends can be pulled out and cut off so that the drill holes can be sealed.

**Representation**

**Properties**

Density: 0,76 kg/cm$^3$;
Type of effect: hydrophobization;
Viscosity: 1,2 mPa.s;
Surface tension: ~ 0,024 N/m.

**Application and limitations**

- It can be applied from inside and/or from the outside of the building materials;
- It can be applied in cases of high degrees of moisture penetration and with all degrees of salt contamination;
- Do not apply in temperatures below 0°C, apply as long as the masonry is not frozen;
- Applied to hollow and crack bricks or hollow masonry, without having to fill the empty brick in advance.

**References : Koster**

**Fig. 1** Example of a HUMITECNIC file – chemical barrier

**Fig. 2** Egas Moniz museum house

**Fig. 3** (**a**) Weakened wooden floor with deformations and (**b**) Micro-organisms and fungal development in the crawlspace

After the rehabilitation of the Egas Moniz museum house in 2008, the ventilation grills were partially closed. Two years after the rehabilitation, the Egas Moniz museum house presented with several pathologies (see Fig. 3a, b), including:

- Weakened wooden floor with deformations;
- Humidity;
- Biological degradation;
- Micro-organisms and fungal development in the crawlspace with high values of relative humidity.

In accordance with the building pathologies observed, the Laboratory of Building Physics (LFC-FEUP) was invited to analyse the causes and propose a treatment methodology. The first step was to monitor the interior conditions (relative humidity and temperature) observed in the crawlspaces below the problematic rooms. Secondly, a numerical analysis was done to support the methodology proposed, with the aim of reducing the moisture level in the air (relative humidity) and maintaining relative indoor [FIX] humidity below 75% (the key to controlling mould growth).

## Experimental Campaign

An experimental campaign was carried out with the continuous monitoring of the relative humidity and temperature in the crawlspaces below certain compartments, as shown in Fig. 4. For this purpose, a great number of data-loggers were placed in strategic locations within the crawlspace (for example, near the ground floor, the wooden floor, the ventilation grills, etc.) to study the influence of relative humidity and temperature on biological degradation of the wooden floor. The data loggers placed near the ventilation grills and in the middle of the crawl space were also able to analyse whether the space was well or poorly ventilated.

Figures 5 and 6 present examples of the relative interior humidity and temperature measured in different compartments and Table 2 shows the statistical values of temperature and relative humidity measured by the different data-loggers.

The results showed that:

- Relative humidity in the crawlspace below the dinner room was approximately 80%;
- In the crawlspace below the salon, the relative humidity measured by data-logger L0.4 was approximately 90% and the outside mean RH was lower than the inside mean RH;
- Relative humidity in the crawlspace below the game room was approximately 93%;
- Relative humidity in the crawlspace below the library was approximately 86%.

In conclusion, after the rehabilitation of the museum house, the ventilation grills were partially closed and the crawlspace was poorly ventilated, essential problems to solve so as to avoid moisture damage. [AU: As originally written, this sentence seemed to indicate that poor ventilation was essential for avoiding moisture damage; I didn't think that was what you were going for, but perhaps I was wrong.] In the literature, it is possible to find different standard building codes specifying different ratios of foundational ventilation areas and crawlspace areas.

## Numerical Simulation

There are an increasing number of computer models that will model heat and mass transfer in building materials [19]. The calculation program used in the numerical simulations for the experimental and analytical validations was WUFI-2D, developed by the Fraunhofer Institute for Building Physics. The governing equation for moisture transport is [21].

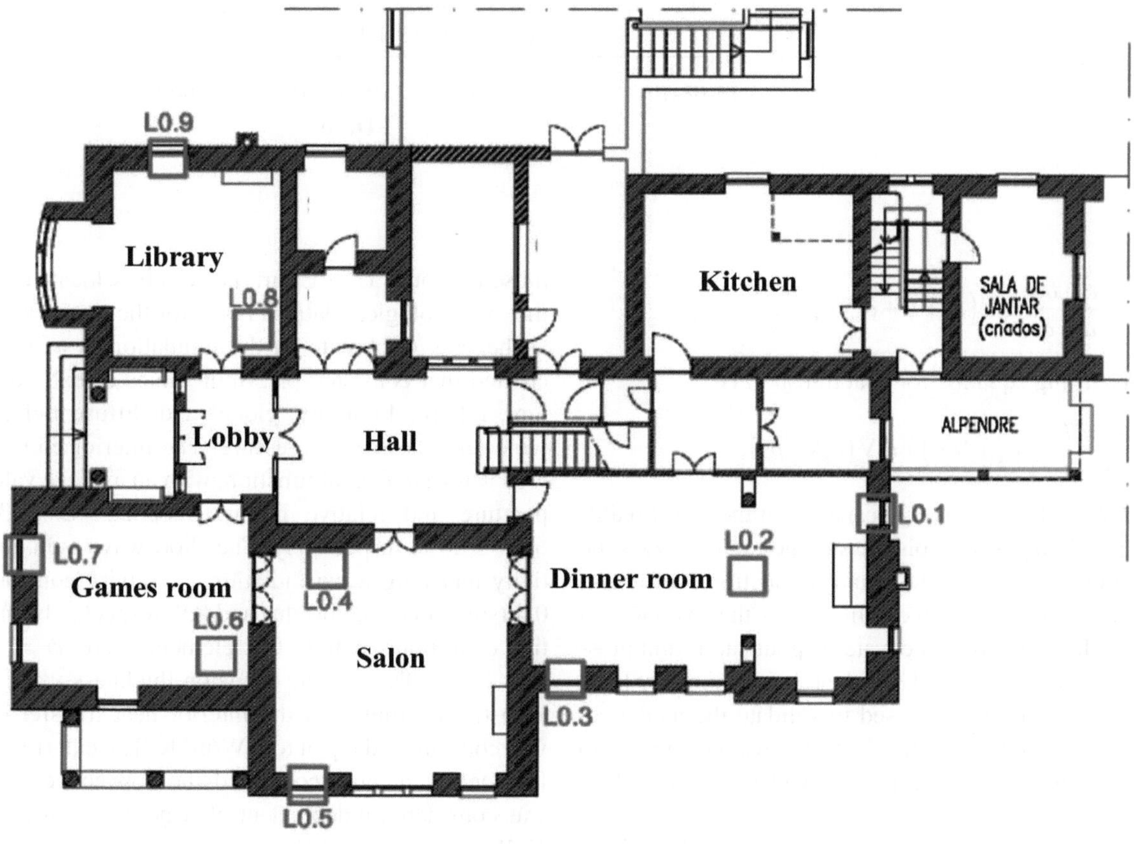


**Fig. 4** Locatio of the data-loggers in the crawlspaces

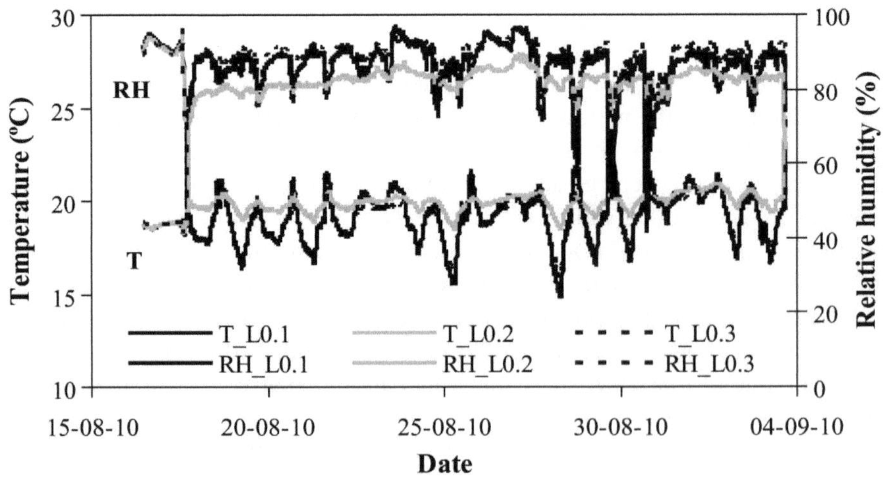


**Fig. 5** Relative interior humidity and temperature (L0.1, L0.2 and L0.3) in the dinner room

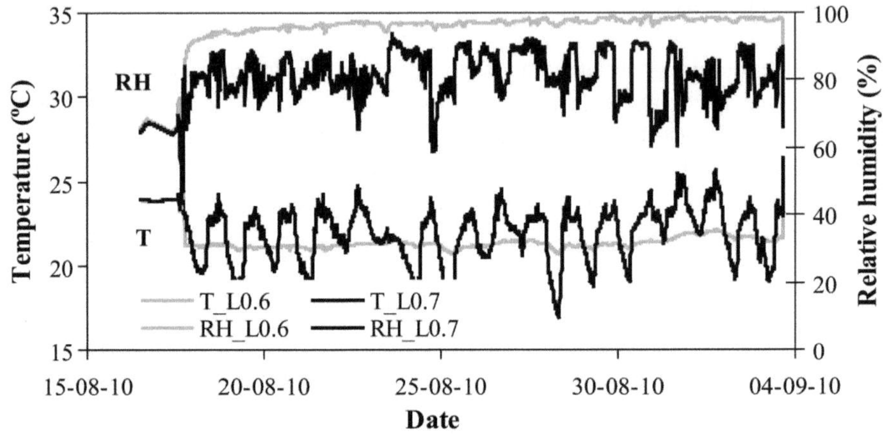


**Fig. 6** Relative interior humidity and temperature (L0.6 and L0.7) in the game room

$$\frac{dw}{d\phi}\frac{\partial \phi}{\partial t} = \nabla\left(D_{\varphi}\nabla\phi + \delta_{p}\nabla\phi p_{a}\right), \quad (1)$$

and the governing equation for heat transport is

$$\frac{dH}{dT}\frac{\partial T}{\partial t} = \nabla\left(\lambda_{T}\nabla T\right) + h_{v}\nabla\left(\delta_{p}\nabla\phi p_{a}\right), \quad (2)$$

where $dH/dT$ is the heat storage capacity of the moist building material, $dw/d\phi$ is the moisture storage capacity, $\lambda_T$ is the thermal conductivity, $D_\phi$ is the liquid conduction coefficient, $\delta_p$ is the water vapour permeability, $h_v$ is the evaporation enthalpy of the water, $p_a$ is the water vapour saturation pressure, $T$ is the temperature and $\phi$ is the relative humidity [10].

In the numerical analysis used to validate the analytical study, only the change in relative humidity was considered to be important, since the experiments took place under isothermal conditions.

The simulation was carried out in time steps of 5 min to reproduce a realistic control. Given that the location of the museum house is in Estarreja, which is located near Porto, the meteorological data of Porto for the year 2005 was used as the exterior climate for the simulation. A weather file was created that consisted of 5-min values of exterior temperature, relative humidity, global and diffuse radiation, rain, wind and barometric pressure. The interior climate considered was a sinusoidal function, with an average value of temperature and relative humidity of 22 °C ± 3 °C and 65% ± 20%, respectively. The short wave radiation absorptivity and long-wave radiation emissivity considered were 0.4 (stucco-normal bright) and 0.9, respectively, and the initial conditions within the element were $\phi$ = 70% and $T$ = 20 °C. The vapour diffusion thickness value used was zero (no coating) and the interior heat transfer coefficient was constant and equal to 8 W/m$^2$ K. The exterior heat transfer coefficient only contained the convective element and was considered independent of wind, at a constant value of 17 W/m$^2$ K (see [7, 13]).

**Table 2** Maximum, minimum and average values of temperature and relative humidity measured by the different data-loggers

| | Outside | | Inside | | L0.1 | | L0.2 | | L0.4 | |
|---|---|---|---|---|---|---|---|---|---|---|
| | T (°C) | RH (%) | T (°C) | RH (%) | T (°C) | RH (%) | T (°C) | RH (%) | T (°C) | RH (%) |
| Mean | 21.8 | 70.7 | 22.5 | 66.9 | 20.5 | 79.6 | 20.5 | 79.6 | 21.4 | 87.7 |
| Min. | 15.3 | 16.7 | 21.4 | 59.0 | 18.5 | 41.2 | 18.5 | 41.2 | 20.3 | 40.9 |
| Max. | 37.6 | 100 | 24.5 | 76.0 | 28.8 | 89.3 | 28.8 | 89.3 | 29.5 | 95.6 |
| | L0.5 | | L0.6 | | L0.7 | | L0.8 | | L0.9 | |
| | T (°C) | RH (%) | T (°C) | RH (%) | T (°C) | RH (%) | T (°C) | RH (%) | T (°C) | RH (%) |
| Mean | 20.3 | 82.2 | 22.2 | 79.1 | 22.2 | 79.1 | 21.8 | 86.2 | 22.3 | 76.3 |
| Min. | 14.5 | 41.8 | 16.9 | 41.4 | 16.9 | 41.4 | 20.1 | 41.4 | 16.6 | 41.5 |
| Max. | 28.7 | 95.0 | 28.9 | 93.7 | 28.9 | 93.7 | 29.0 | 93.9 | 29.5 | 92.2 |

**Table 3** Material properties of Limestone

| | | | |
|---|---|---|---|
| Bulk density, $\rho$ (kg/m³) | 2155 | | |
| Heat capacity, $c_p$ (J/kgK) | 1000 | | |
| Porosity, $\varepsilon$ (%) | 19.7 | | |
| Thermal conductivity, $\lambda_T$ (W/mK) | 1.33 | | |
| Vapour diffusion resistance factor, $\mu$ (−) | Dry cup:<br>Wet cup: | 41<br>29 | |
| Moisture storage function, $w$ (kg/m³) | $\phi$ (%) | *Adsor* | *Desadsor* |
| | 4.0<br>11.2<br>34.8<br>58.6<br>76.3<br>80.0<br>84.2<br>92.1 | 0.521<br>0.593<br>0.872<br>1.043<br>1.237<br>1.334<br>1.584<br>2.831 | 0.951<br>1.150<br>1.239<br>1.628<br>2.360<br>–<br>–<br>4.144 |
| Capillary transport coefficient, $D_w$ (m²/s) | $w$ = 1.7 kg/m³<br>$w$ = 188 kg/m³ | $6.6 \times 10^{-11}$<br>$6.2 \times 10^{-8}$ | |
| Water absorption coefficient, $A$ (kg/m²s$^{1/2}$) | 0.024 | | |
| Free-water saturation, $w_f$ (kg/m³) | 177 | | |

The simulations used the real climatic data and actual material properties of the Egas Moniz museum house (see Table 3). Detailed model geometry of the game room was created in WUFI-2D (see Fig. 7) to be used as an example of a numerical study.

## Proposed Intervention Methodology – The Hygro-regulated Device

Reducing the moisture level in the air (relative humidity) and maintaining a relative inside humidity below 75% is the key to controlling mould growth. The numerical results showed that an unventilated crawlspace or one that is continuously ventilated, through the increase in the air exchange rate, has biological degradation. Increasing the air exchange rate from a certain level only reduces the humidity if the ventilation removes more moisture than is added by increased ground moisture evaporation. In accordance with this, a hygro-regulated system, with the air extraction controlled by a variable ventilator, which comes into operation when the inside water vapour pressure is higher than the outside water vapour pressure, is the solution proposed.

The intervention methodology proposed could be divided into two steps: control the groundwater level and ventilate the crawlspaces with a hygro-regulated system [11, 14]. The procedure adopted for the crawlspaces was the implementation of a hygro-regulated ventilation system with the following characteristics:

- A ventilation network is created, by a crawlspace module, consisting of a pipe 0.075 m in diameter associated with a hygro-regulated system (Fig. 8);
- The ventilation pipes should be individual, with different flow regulations, and all the pipes should then converge into a single pipe, where the temperature and relative humidity of the crawlspace's air is measured;
- The hygro-regulated device receives records related to the inside and outside hygrothermal conditions, and comes into operation only when the outside water vapour pressure is lower than the partial pressure inside of the crawlspace;
- The ventilator should have variable speed, with flows between 10 and 50 m³/h.

In this novel device, the extraction is controlled by a variable ventilator, hygro-regulated, which comes into operation when the inside water vapour pressure is higher than the outside water vapour pressure, and when the inside relative humidity is higher than a certain predefined value so as to guarantee that no salt crystallization/dissolution problems occur inside the system (see Fig. 9). The control module receives information from two probes (inside and outside temperature and relative humidity), calculates the inside and outside water vapour pressure and evaluates the positive or negative sign of the pressure differential for shutting the ventilator on or off. The data acquisition system stores the data received by probes, for evaluation of this solution's effectiveness.

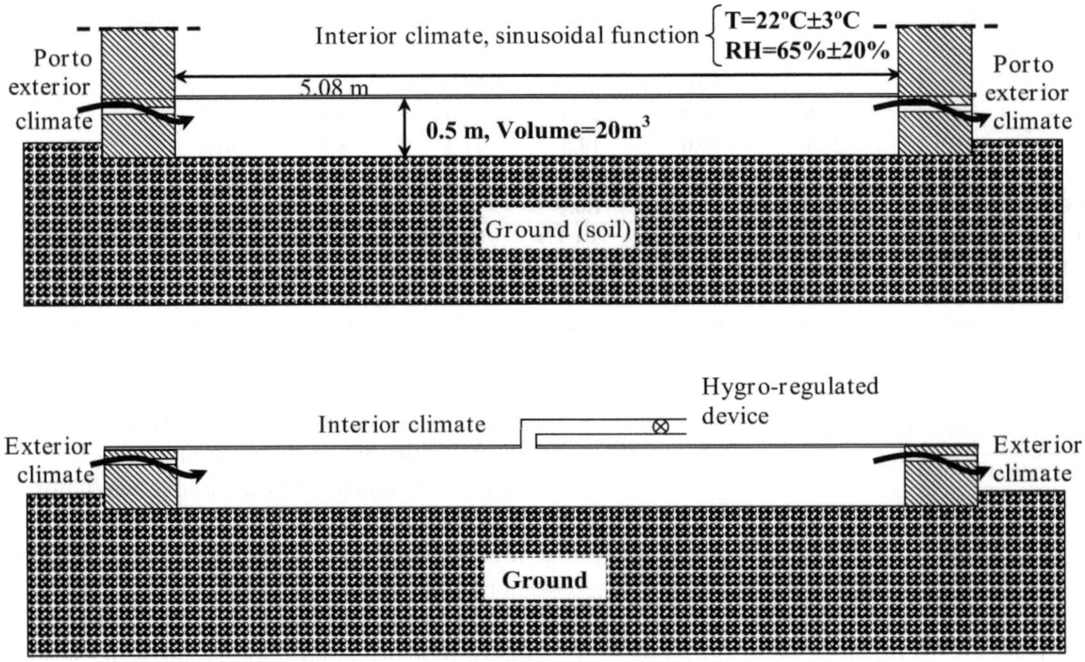


**Fig. 7** Detailed geometry, created with WUFI-2D, of the analysed building area (game room)

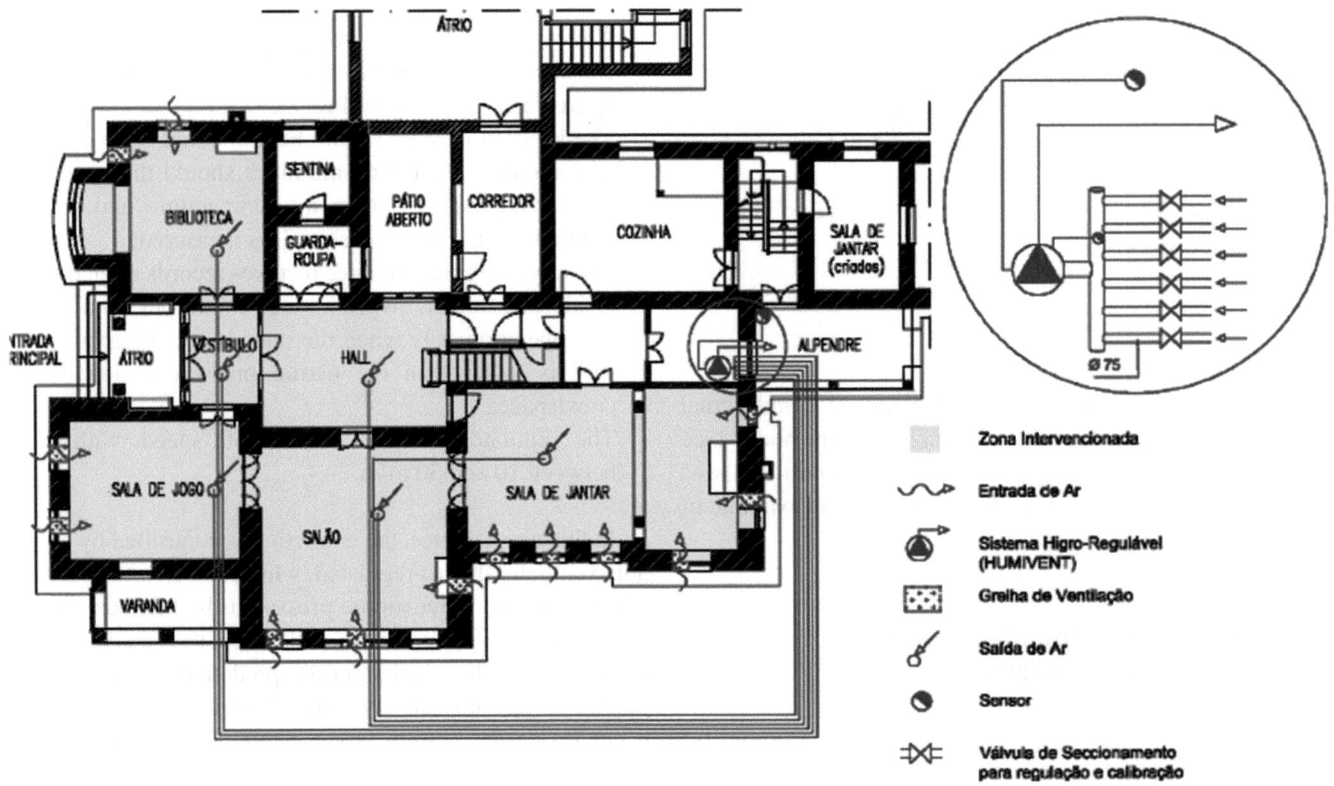


**Fig. 8** Scheme of the intervention methodology proposed for the crawlspaces

**Fig. 9** The principle of functioning of the hygro-regulated ventilation system

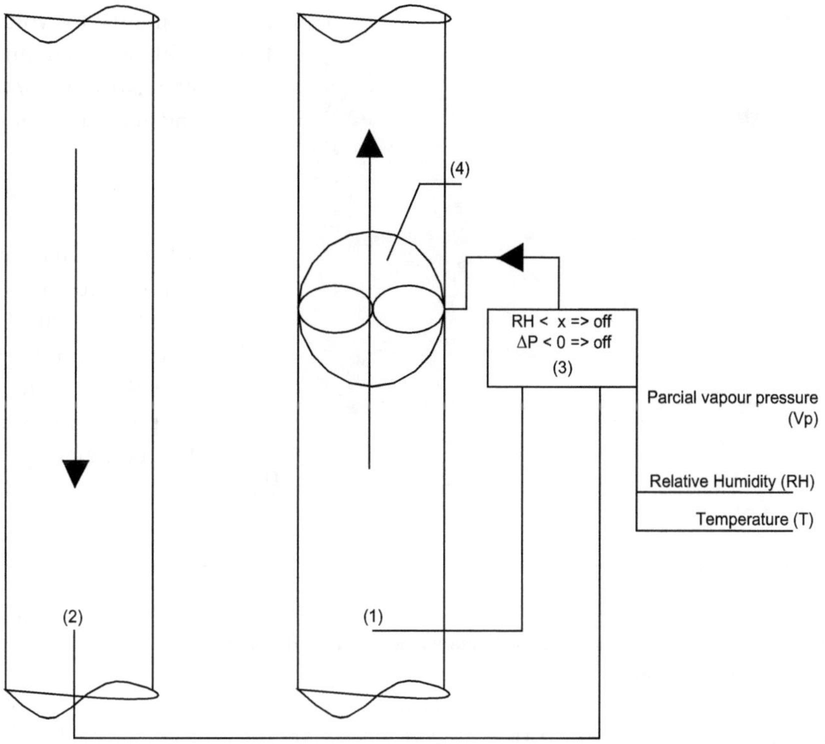


**Legend:**

Probe (1) – Entrance (temperature and relative humidity); Probe (2) – Exit (temperature and relative humidity); Fan with variable speed (3)

Control module (4): Probe 1 – $T_1$ e $RH_1$ => Vapour pressure 1 (Pa)
Probe 2 – $T_2$ e $RH_2$ => Vapour pressure 2 (Pa)
Calculate $\Delta P = P_2 - P_1$

Ventilator operation: ON if $\Delta P \geq 0$ and $RH_1$ > Predetermined value
OFF if $\Delta P \leq 0$ or $RH_1$ < Predetermined value

The hygro-regulated ventilation system presented is an adaptation of the "Hygro-Regulated Wall Base Ventilation System for the Treatment of Rising Damp" developed by the Laboratory of Building Physics, LFC-FEUP (see [32]).

The prototype produced and tested is composed of a programmer data-logger linked to two relative humidity (*RH*) and temperature (*T*) probes, which are themselves linked up to one to four ventilators at the same time, a battery, and a computer that programs the operational device and gives instructions to the data-logger to provide temporal *RH* and *T* information, a reading of the records and data processing (see Fig. 10).

## Mould Growth Prediction

### Analytical Analyse

In order to predict mould growth, the equation developed by Viitanen [31] for wood (specifically pine and spruce) was used. The criterion presented by the author for avoiding mould growth on wood is that *t* must be shorter than the time of wetness (TOW), i.e., TOW < < *t*, with *t* given by

$$t = \exp\left(-0.68 \ln T - 13.9 \ln RH + 0.14W - 0.33S + 66.02\right), \quad (3)$$

where *t* is the time for mould growth to reach the microscopic stage (weeks), *T* is the temperature (0.1–40 °C), *RH* is the relative humidity (%), *W* is the wood species (pine = 0, spruce = 1) and *S* is the factor describing the nutrients on the wooden surface (a re-sawed surface after drying = 0, a kiln-dried surface = 1).

TOW is defined here as the cumulative time period when *RH* is higher than 75%. This means that the counting of TOW starts when *RH* exceeds 75%. The periods when *RH* < 75% are not taken into account. The results presented in Table 4 show a high prediction of mould growth in the game room (L0.6), TOW > *t*, and a significant possibility of mould growth in the salon (L0.4) and library (L0.8),

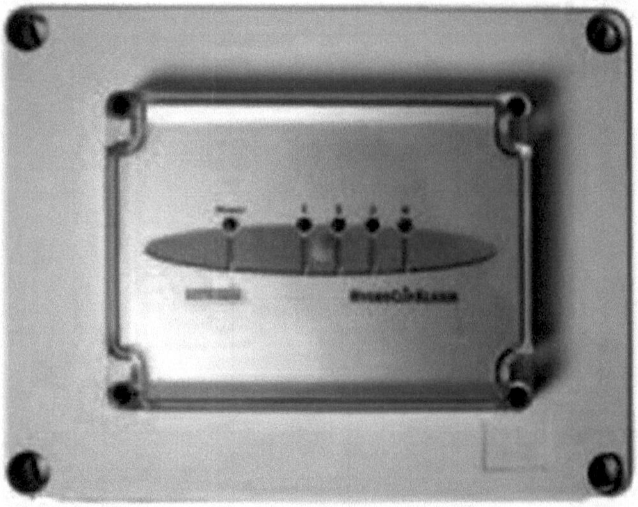

**Fig. 10** The prototype developed

**Table 4** Values of time for mould growth and wetness obtained in the Egas Moniz museum house

| | L0.2 | L0.4 | L0.6 | L0.8 |
|---|---|---|---|---|
| TOW (weeks) | 2.39 | 2.41 | 2.41 | 2.51 |
| *t* (weeks) | 10.26 | 2.50 | 1.01 | 2.83 |

TOW < *t*. The only room analysed with a lower prediction of mould growth for the period studied is the dining room (L0.2).

In the crawlspace, it is usually considered that a relative humidity over 75%–80% [30], during a period of several weeks or months, can cause mould growth. Temperature is high enough for mould growth, because it is usually over +5 °C in crawlspaces. Relative humidity in the crawlspace is the result of ground moisture evaporation, air change rate and thermal behaviour, all of which are strongly linked.

In a general state, the humidity balance for the air in the crawlspace is given by [8, 24]:

$$V\frac{\partial C_{air}}{\partial t} = g_{vent}^{in} + g - g_{vent}^{out}, \tag{4}$$

where $V$ is the volume of the crawlspace (m$^3$), $g$ is the moisture flow (kg/s) and $C_{air}$ is the humidity by volume of crawlspace air (kg/m$^3$).

If the ground moisture evaporation and moisture flows carried by air change are the only moisture flows in the crawlspace, i.e., moisture storage and flows in constructions are not considered, then rain and surface water are not present in the crawlspace, as it should be, provided the design is correct. Therefore, the moisture balance can be written in steady state with the r.h.s. equal to zero, or

$$C_{air}^{out} q_m = C_{air} q_m + g, \tag{5}$$

where $C_{air}^{out}$ is the outdoor air humidity by volume (kg/m$^3$), $q_m$ is the air change in the crawlspace (m$^3$/s) and $C_{air}$ is the humidity by volume in the extracted air (considering that, with complete mixing, the air is the same everywhere in the crawlspace) (kg/m$^3$). Ground moisture evaporation based on the mass transfer coefficient and potential is

$$g = \beta\left(p_{ground} - p_{air}\right)\frac{M_w}{R.T}A_{evap}, \tag{6}$$

where $\beta$ is the mass transfer coefficient (m/s), $p_{ground}$ is the vapour pressure on the ground surface (Pa) and $p_{air}$ in the crawlspace air (Pa), $M_w$ is the molecular weight of water (0.018 kg/mol), $R$ is the universal gas constant (8.31 J/mol K), $T$ is the absolute temperature (K) and $A_{evap}$ is the evaporation surface area (m$^2$). To determine the mass transfer coefficient, $\beta$, the following equations could be used:

$$\beta = \frac{2.2\left(T_{ground} - T_{air}\right)^{1/3}}{\rho.c_p} \quad \left(\text{laminar flow, ach} < 1.5\text{h}^{-1}\right), \tag{7}$$

$$\beta \approx \frac{2.2\left(T_{ground} - T_{air}\right)^{1/3} + 5.4q}{\rho.c_p} \quad \left(\text{turbulent flow, ach} > 1.5\text{h}^{-1}\right), \tag{8}$$

where $\rho$ is the density of air (kg/m$^3$), $c_p$ is the specific heat capacity of air (J/kg K), $T_{ground}$ is the temperature on the ground surface, $T_{air}$ is the temperature of the crawlspace air and $q$ is the airflow (m$^3$/s).

However, it is important to keep in mind that if the ground soil is relatively dry, Eq. (6) overestimates the evaporation rate and, alternatively, Fick's law should be used for moisture transfer inside the ground surface. In our case study, for the time period under analysis (one month), and using Eq. (5), the average value of moisture evaporation was approximately 0.21 g/m$^2$ h (natural ventilation of 0.1 ach), corresponding to a mass transfer coefficient, $\beta$, of approximately 6.6 × 10$^{-5}$ m/s. The development of Eq. (4) results in

$$V\frac{\partial C_{air}}{\partial t} = achV\left(C_{air}^{out} - C_{air}\right) + \beta\frac{V}{L}\left(C_{ground} - C_{air}\right), \tag{9}$$

where $L$ is the height of the crawlspace. The integration of Eq. (9) for the initial and end conditions $t = 0$ to $t = t$ and $C = C_0$ to $C = C$ gives us

$$C_{air} = \frac{achC_{air}^{out} + \frac{\beta}{L}C_{ground}}{ach + \frac{\beta}{L}} - \frac{e^{-t\left(ach+\frac{\beta}{L}\right)+\ln\left[ach\left(C_{air}^{out}-C_0\right)+\frac{\beta}{L}\left(C_{ground}-C_0\right)\right]}}{ach + \frac{\beta}{L}}. \tag{10}$$

For $t \to \infty$, steady-state, Eq. (11) results in

$$C_{air} = \frac{achC_{air}^{out} + \frac{\beta}{L}C_{ground}}{ach + \frac{\beta}{L}}. \quad (11)$$

So, for the game room crawlspace ($A_{evap}$ = 40m² and $L$ = 0.5 m) with an $ach$ = 0.1h$^{-1}$, and considering an average value of temperature and relative humidity of 15 °C and 77% for Porto, and $C_{ground}$ = 0.013 kg/m³ (corresponding to a value of relative humidity of 97%, practically fully saturated), it results, according to Eq. (11), in a $C_{air}$ of 0.012 kg/m³, corresponding to a value of relative humidity of 94%. Increasing the $ach$ value to 0.5 h$^{-1}$ (10 m³/h) and 2.5 h$^{-1}$ (50 m³/h) results, according to Eq. (11), in a relative humidity value in the crawlspace of 87% and 80%, respectively.

It is possible to observe that the value obtained with Eq. (11), $C_{air}$ = 0.012kg/m³ and $RH$ = 94%, for the case of crawlspace with natural ventilation is practically the same as that obtained in in-situ experiments, $RH$ = 93% (L0.6). The results also show that the use of natural ventilation in this crawlspace is insufficient to avoid mould growth ($RH$ < 80%).

## Numerical Analyse

In this case study, simulations were done for 1 year with a real exterior climate so as to analyse the influence of ventilation on a crawlspace with wet ground, practically fully saturated (≈100% RH). The following figures show the importance of good ventilation in crawlspaces:

In Figs. 11 and 12, the numerical simulation results for the same crawl space, but with ventilation, $ach$ = 0.1 h$^{-1}$ and $ach$ = 2.5 h$^{-1}$, are presented. In Fig. 11, it is possible to observe the relative humidity in the crawlspace near the wooden floor and in the middle of the wooden floor. This figure shows an extremely high tendency for a crawlspace in a moderate climate without ventilation to present mould growth (during the period from February to August). The use of mechanical ventilation with an $ach$ of 2.5 h$^{-1}$ (see Fig. 12) decreases the relative humidity, but does not solve the problem of mould growth. In conclusion, this case study shows that ventilation is needed to avoid mould growth in crawlspaces. However, it is preferable to use controlled mechanical ventilation, such as a hygro-regulable mechanical ventilation device, that becomes operational only when the outside water vapour pressure is lower than the partial pressure inside the crawlspace. We believe that a hygro-regulable system is a good solution for this kind of mould problem in crawlspaces [15].

## Conclusions

Ground moisture evaporation is a key element in moisture balance in crawlspaces. If any moisture evaporation occurs, ventilation will always be required to remove this moisture. However, increasing the air change rate from a certain level only reduces the humidity if the ventilation removes more moisture than is added by increased ground moisture evaporation.

The hygro-regulated device is presented as a good solution for avoiding high levels of moisture in crawlspaces. This technique consists in extracting the air in the crawlspace through the installation of a hygro-regulable mechanical ventilation device. It was successfully tested in the treatment of the rising damp phenomenon in historical buildings, and future experimental results will be used to validate this device in crawlspaces.

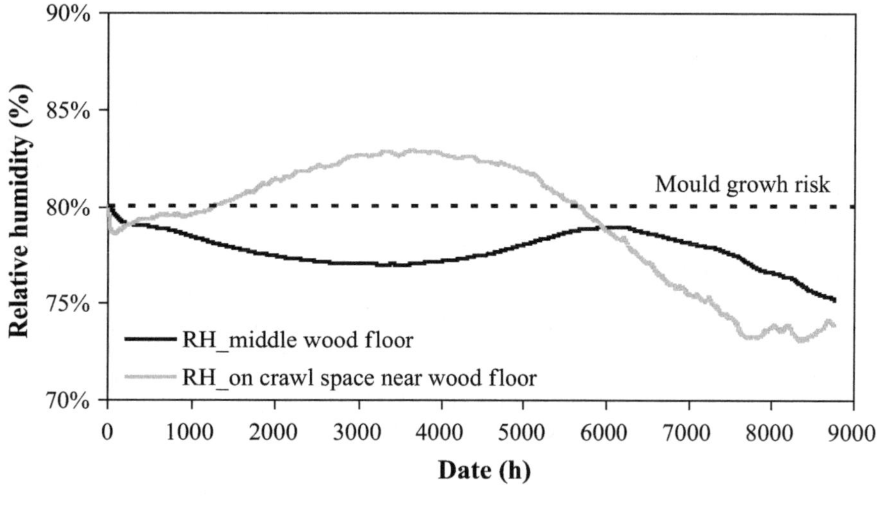


**Fig. 11** Values of relative humidity in a crawlspace with minimal ventilation (ach = 0.1 h$^{-1}$)

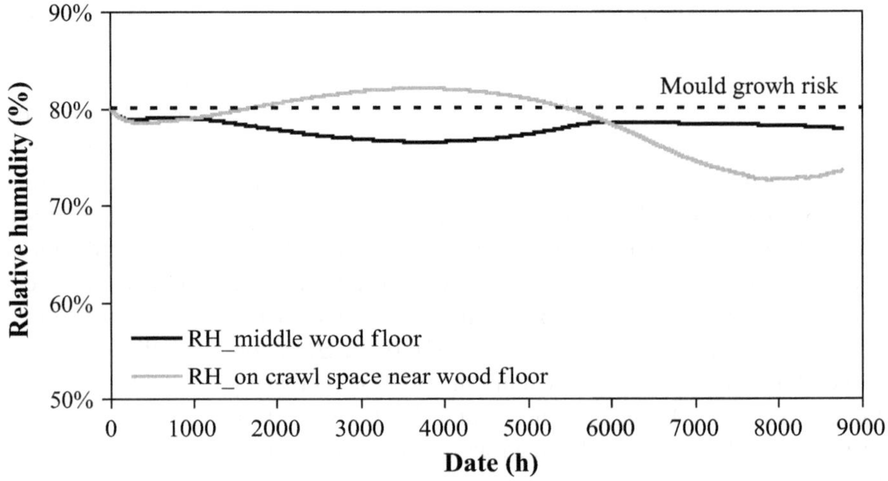


**Fig. 12** Values of relative humidity in a crawlspace with adequate ventilation (ach = 2.5 $h^{-1}$)

## References

1. Bakolas A, Biscontin G, Moropoulou A, Zendri E (1996) Salt impact on brickwork along the canals of Venice. Mater Struct 29(1):47–55. https://doi.org/10.1007/BF02486006
2. Block SS (1953) Humidity requirements for mould growth. Appl Microbiol 1(6):287–293
3. Bowley M (1975) Desalination of stone: a case study, Paper CP46/75. Building Research Establishment, Watford, UK
4. Burke S (2005) Advantages and risks associated with crawl space foundations. In: Proceedings of the 7th symposium on building physics in the nordic countries, June 13–15, Reykjavík Iceland
5. Burke S (2007) Crawl spaces in wood framed single family dwellings in Sweden: unwanted yet popular. Struct Surv 25(1):51–60
6. Cunningham MJ, Roos C, Gu L, Spolek G (2004) Predicting psychrometric conditions in biocontaminant microenvironments with a microclimate heat and moisture transfer model – description and field comparison. Indoor Air 14(4):235–242
7. Delgado JMPQ, Ramos NMM, Barreira E, Freitas VP (2010) A critical review of hygrothermal models used in porous building materials. J Porous Media 13(3):221–234
8. Elmroth A (1975) Crawl space foundation. Construction Research, Report 12, Stockholm
9. Emenius G, Svartengren M, Korsgaard J, Nordvall L, Pershagen G, Wickman M (2004) Building characteristics, indoor air quality and recurrent wheezing in very young children (BAMSE). Indoor Air 14(1):34–42
10. Evrard A, De Herde A (2010) Hygrothermal performance of Lime-Hemp wall assemblies. J Build Phys 34(1):5–25
11. Freitas VP, Guimarães AS, Delgado JMPQ (2011) The HUMIVENT device for rising damp treatment. Recent Patents Eng 5(3):233–240
12. Guimarães AS, Delgado JMPQ, Freitas VP (2014) Rising damp: optimization of the wall base ventilation system. Defect Diffus Forum 353:311–316
13. Guimarães AS, Delgado JMPQ, Freitas VP (2013) Rising damp in walls: evaluation of the level achieved by the damp front. J Build Phys in Press
14. Guimarães AS, Delgado JMPQ, Freitas VP (2010) Mathematical analysis of the evaporative process of a new technological treatment of rising damp in historic buildings. Build Environ 45(11):2414–2420
15. Hagentoft CE, Kalagasidis AS, Nilsson SF, Thorin M (2008) Mould growth control in cold attics through adaptive ventilation. In: Proceedings of the 8th Nordic symposium on building physics, Copenhagen, pp 1–8
16. Haverinen U, Husman T, Pekkanen J, Vahteristo M, Moschandreas D, Nevalainen A (2001a) Characteristics of moisture damage in houses and their association with self-reported symptoms of the occupants. Indoor Built Environ 10(2):83–94
17. Haverinen U, Husman T, Vahteristo M, Koskinen O, Moschandreas D, Nevalainen A, Juha P (2001b) Comparison of two-level and three-level classifications of moisture-damaged dwellings in relation to health effects. Indoor Air 11(3):192–199
18. Henningsson B (1980) Thermotolerant moulds on timber during kiln drying. Uppsala, Swedish University of Agricultural Sciences, Department of Forest Products, Note No 96
19. Hens H (1996) Final report task 1: modelling common exercises, Summary reports. Annex 24
20. Hinks J, Cook G (1999) The technology of building defects. E & FN Spon, London
21. Holm AH, Kunzel HM (2003) Two-dimensional transient heat and moisture simulations of rising damp with WUFI-2D. In: Proceedings of the 2nd international conference on building physics, Leuven, Belgium, 14–18 September 2003, Swets & Zeitlinger B.V., Lisse, pp 363–367
22. Iwamae A, Matsumoto M (2003) The humidity variation in crawl spaces of Japanese houses. J Therm Envel Build Sci 27(2):123–133
23. Iwamae A, Suzuki H (2008) Durability of crawl space based on damage due to wood rot. In: Proceedings of the 8th Nordic symposium on building physics, Copenhagen, pp 747–754
24. Kurnitski J (2001) Ground moisture evaporation in crawl spaces. Build Environ 36(3):359–373
25. Kurnitski J, Matilainen M (2000) Moisture conditions of outdoor air-ventilated crawl spaces in apartment buildings in a cold climate. Energ Build 33(1):15–29
26. Matilainen M, Pasanen P (2002) Transport of fungal spores from crawl space to indoors. In: Indoor Air Monterey, USA

27. Matilainen M, Kurnitski J, Seppänen O (2003) Moisture conditions and energy consumption in heated crawl spaces in cold climates. Energ Buildings 35(2):203–216
28. Nofal M, Kumaran K (2000) On implementing experimental biological damage-functions models in durability assessment system. In: Proceedings of Japan-Canada housing R&D experts working group meeting building envelope, pp 111–124
29. Park D (1982) Phylloplane fungi: tolerance of hyphal tips to drying. Trans Br Mycol Soc 79(1):174–179
30. Samuelsson I (1994) Moisture control in crawl space. ASHRAE Tech Data Bull 10(3):58–64
31. Viitanen H (1996) Factors affecting the development of mould and brown rot decay in wooden material and wooden structures: Effect of humidity, temperature and exposure time. Dissertation, Uppsala
32. WIPO (2011) Hygro-adjustable wall base ventilation system for the treatment of rising damp. Publication n. WO/2010/093272, University of Porto. Webpage: https://patentscope.wipo.int/search/en/detail.jsf?docId=WO2010093272. Accessed in 6 of November 2017

MIX
Papier aus verantwortungsvollen Quellen
Paper from responsible sources
FSC® C105338

If you have any concerns about our products,
you can contact us on
**ProductSafety@springernature.com**

In case Publisher is established outside the EU,
the EU authorized representative is:
**Springer Nature Customer Service Center GmbH**
**Europaplatz 3, 69115 Heidelberg, Germany**

Printed by Libri Plureos GmbH
in Hamburg, Germany